Design Research Foundations

The goal of the series is to provide a platform for publishing state of the art research on foundational issues in design and its applications in industry and society. Suitable topics range from methodological issues in design research to philosophical reflections on the specificities of design rather than actual design work or empirical cases only. The definition of design behind the series is inclusive. In terms of disciplines, it ranges from engineering to architecture. In terms of design work, it ranges from conceptual issues in design through design experiments and prototypes to evaluative studies of design and its foundations.

Proposals should include:

- A proposal form, as can be found on this page
- A short synopsis of the work or the introduction chapter
- The proposed Table of Contents
- The CV of the lead author(s)
- If available: one sample chapter

We aim to make a first decision within 1 month of submission. In case of a positive first decision the work will be provisionally contracted: the final decision about publication will depend upon the result of the anonymous peer review of the complete manuscript. The series editors aim to have the complete work peer-reviewed within 3 months of submission.

The series discourages the submission of manuscripts that contain reprints of previous published material and/or manuscripts that are below 150 pages / 75,000 words.

For inquiries and submission of proposals authors can contact the series editors, Ilpo Koskinen via: ilpo.koskinen@polyu.edu.hk or Peter Gall Krogh via: pkrogh@cc.au.dk

Shunying An Blevis · Eli Blevis · Makayla Lewis ·
Miriam Sturdee
Editors

Seeing Interaction Design

Springer

Editors
Shunying An Blevis
Luddy School of Informatics, Computing, and Engineering
Indiana University Bloomington
Bloomington, IN, USA

Eli Blevis
Luddy School of Informatics, Computing, and Engineering
Indiana University Bloomington
Bloomington, IN, USA

Makayla Lewis
Networks and Digital Media
Kingston University
London, UK

Miriam Sturdee
School of Computer Science
University of St Andrews
North Haugh, UK

ISSN 2366-4622 ISSN 2366-4630 (electronic)
Design Research Foundations
ISBN 978-3-032-14199-6 ISBN 978-3-032-14200-9 (eBook)
https://doi.org/10.1007/978-3-032-14200-9

This Springer imprint is published by the registered company Springer Nature Switzerland AG
The registered company address is: Gewerbestrasse 11, 6330 Cham, Switzerland

Acknowledgments

Heartfelt thanks to the series editors and others involved in the production of this book. Heartfelt thanks to the many and various people who have contributed to the acceptance of pictorials and visual forms within HCI and Design.

Contents

Chapter 1
Introduction

Shunying An Blevis, Eli Blevis, Makayla Lewis, and Miriam Sturdee

> "Even though our relation to experience is often (and increasingly) mediated by visual formats and images, the bias against visual forms of knowledge production is longstanding in our culture. Logocentric and numerocentric attitudes prevail."
>
> —Johanna Drucker (2014): 16

1.1 Introduction

The *Seeing* in *Seeing Interaction Design* refers to the idea that seeing is an act of knowing—that is to say, we hold that ways of seeing & ways of showing *are* ways of knowing & ways of saying. The chapters of this book embrace and develop these notions as an alternative to prevailing notions in interaction design and HCI which emphasize texts over visual forms.

S. A. Blevis · E. Blevis (✉)
Luddy School of Informatics, Computing, and Engineering, Indiana University at Bloomington, Bloomington, USA
e-mail: eblevis@iu.edu

S. A. Blevis
e-mail: san@iu.edu

M. Lewis
Kingston University, London, UK
e-mail: m.m.lewis@kingston.ac.uk

M. Sturdee (✉)
University of St, Andrews, St. Andrews, Scotland
e-mail: ms535@st-andrews.ac.uk

S. An Blevis et al. (eds.), *Seeing Interaction Design*, Design Research Foundations,
https://doi.org/10.1007/978-3-032-14200-9_1

This introduction is organized in the following way:

1. First, we describe the organization of the book as a whole
2. We briefly clarify a few important points prompted by discussions with others about the book—*a few things, first*
3. We give an account of the origin of pictorials as a form
4. We classify the pictorial chapters of the book in terms of their means of production, namely *By Eye*, *By Hand*, *By Prompt*, and *By Speculation*
5. We consider if visual forms can be or can create knowledge
6. We give a summary and some additional details—*a few things, finally*—ahead of the chapter summaries
7. We give a short summary of each chapter, together with a selected thumbnail to represent the style of each, and a classification in terms of means of production

1.2 About This Book

Our book brings visual perspectives to interaction design, broadly defined. These take the form of pictorial chapters. Pictorials are essays in which the visual materials are the most important contributions. Pictorials are a rigorously and competitively reviewed scholarly form that has become commonplace in some HCI and some Design venues over this past decade. These venues include Designing Interactive Systems DIS (since 2014), Creativity and Cognition C&C (since 2017), Tangible and Embedded Interaction TEI (since 2020), International Association of Societies of Design Research IASDR (since 2021). In these contexts, Pictorials have been treated with the same rigorous reviewing standards as papers. We return to *the origin story of pictorials* in the section of the same name, below.

The four contributing editors and authors have all participated, along with many others, in the genesis of these forms and their acceptance into the scholarly canon.

The book is unusual in the sense that it is half monograph and half anthology—a *monthology*. The intent of the book is to archive a curated selection of years of contributions to pictorials in various venues with added commentary.

We have invited several others to contribute a pictorial chapter for the anthology portion of this book. We asked authors who have made (or in some cases can make) important contributions to pictorials germane to this volume that connects pictorials and visual ways of knowing to interaction design, broadly defined. Many of the pictorial chapters are based on work which has already appeared in conference proceedings and represent a curation of some of the best work which has appeared.

The book is organized into five main sections, namely:

1. Sustainability
2. Seeing and Noticing
3. Pushing Boundaries
4. Inspirations
5. For Educators

Each section contains 3–4 pictorial chapters and 1 or 2 pictorial chapters of each of these sections are by some combination of the contributing editors or co-authors including the contributing editors.

These sections are not the only categories which may be imagined for pictorial contributions. They owe to an exegesis classification of our own prior work. The invitations to others were in turn based on this classification. This means that there are many fine pictorials in the canon that merit publication in the form this book takes. In compiling this present volume, we realized that there may be many additional volumes along these lines with their own exegetical classifications and curations. We hope to have a hand in compiling these as well in the future and welcome others to do so.

1.3 A Few Things, First

1.3.1 Wandering Around

To make things very clear, the five major sections of the book are not exhaustive nor are they intended to be. Think of this volume as a curation of five topics we think are of interest. Like an Art gallery has curated exhibits (see Fig. 1.1: Wandering Around [GAI]), expect that there are many other thematic curations on different topics to appear. The five sections and fifteen chapters represent some points within the affine space of pictorials in which the reader/viewer is invited to *wander around*. Below, we classify each chapter in terms of their means of production. In the final section of this introduction, we summarize each chapter.

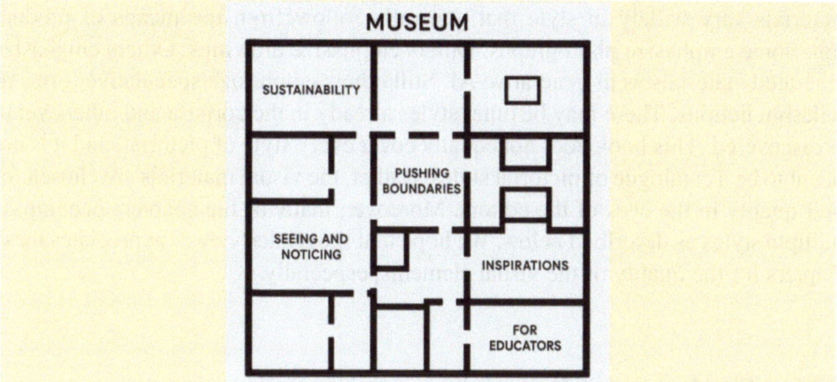

Fig. 1.1 Wandering Around [GAI]. *ChatGPT. Prompt: Please draw the floorplan of a museum with 10 exhibits, with 5 of the exhibits named (1) sustainability (2) seeing and noticing (3) pushing boundaries (4) inspirations and (5) for educators*

1.3.2 Forms of Knowledge

In all that follows, we argue that visual materials—images and drawings—are important in and of their own right. We argue that visual materials are objects of knowledge as much as texts are objects of knowledge. We argue that seeing is an act of knowing as much as reading is an act of knowing. Not everyone agrees—this book is no more the last word on this subject than it is the last curation of images and drawings of interest that will ever be.

1.3.3 Meaning and Layout

In the decade plus old corpora of pictorials in the ACM and IASDR venues enumerated above, typographic layout is considered by many to be part of the contributions under review. The result is that the pictorials in these corpora are quite varied in style and appear quite differently as a matter of presentation. Ideas about the importance of layout can vary as well, from layout serving as an essential part of the expression to layout serving best when it distracts least from the actual images or drawings themselves. It is the latter perspective that informs this book. Note that these are not mutually exclusive criteria. That is, a simple layout that lets the visual materials stand out without the distraction of any decorative elements can be a form of intentional layout.

1.3.4 Pictorial Styles

Pictorials vary widely in style that generally follow from the means of production. Some emphasize photographs. Others emphasize drawings. Others emphasize generated materials as in generative AI. Still others emphasize speculative forms as in design fictions. There may be other styles already in the corpora and others yet to be discovered. This book does not equally cover every style of pictorial, and it is not meant to be a catalogue of pictorial styles. Rather, the visual materials are chosen for their quality in the eyes of the editors. Moreover, many of the chapters encompass multiple styles as described below. We hope that the reader/viewer appreciates these chapters for the quality of the visual elements, especially.

1.3.5 The Future of Pictorials

One would not expect to write a book, whether a monograph, anthology, or monthology as in the volume at hand—and further expect that volume to prescribe the future of books. Similarly, this book does not strictly predict the future of pictorials. One may expect that pictorials as a form will continue to be accepted and evolve in the context of proceedings venues—it is already a big accomplishment that pictorials are accepted as archival contributions in some venues in the same way as papers are. The best work of these proceedings warrants presentation in book form such as the current volume. Such a book form, in our view, can better emphasize the quality of the visual materials over any other considerations. We expect that such book forms can evolve together with an evolving notion of pictorials.

1.4 The Origin Story of Pictorials

1.4.1 An Historical Account

The first pictorial to be accepted into the HCI archive was a 2012 NordiCHI pictorial paper by Nadine Jarvis, David Cameron, and Andy Boucher, titled *Attention to detail: annotations of a design process.* Boucher revisits this work and contextualizes its contributions in the third chapter of the *Inspirations* section of this volume.

In 2014, working with William Odom and Sabrina Hauser, Blevis published an account of the origin story of pictorials in ACM *interactions* at the invitation of Odom and Hauser (Blevis et al., 2014). In that article, the images and permissions to use them were curated and secured by Odom and Hauser and the texts were written by Blevis. The first pictorials track in an ACM conference was organized by Nadine Jarvis, David Cameron, Hauser, and Odom. Blevis submitted a pictorial to that track that was selected as best pictorial (Blevis, 2014). The pictorial is entirely photographic. Blevis was invited by Hauser and Odom to co-author the *interactions* article because of this pictorial. We reproduce in quotes the account of the origins of pictorials that appeared in the *interactions* article below with some minor editing. Of course, other accounts are possible and indeed likely. We present this account as a starting point for discussion, rather than an *ex-cathedra* statement of how things actually were.

> *"The idea of images as a foundational mode of creating and articulating knowledge about interactivity has been gaining traction in HCI and interaction design. The use of photography and imagery has long been foundational in traditional schools of design. Inventories of "the best" such schools—however subjective—can be found in many business press sources. Here, we privilege two design schools as exemplars of these traditions, owing to their associations with two key figures, László Moholy-Nagy and William Gaver. Famous for its history in connecting photography and design, the Institute of Design in Chicago was founded some eighty-five plus years ago by painter and photographer Lászlό Moholy-Nagy from the German Bauhaus. In HCI the centrality of visual form had been advanced notably by Goldsmiths at the University of London under the stewardship of William Gaver, together with Andy Boucher whose account of pictorials appears later in this volume. Privileging these two institutions here must be accompanied by the awareness that many other fine schools and distinguished figures in the design tradition were equally important in establishing this foundational role for photography and imagery.*

The "Visual Thinking Gallery" (now titled "Exit") that has appeared on the inside back page of each issue of Interactions since September 2011 presents a photograph relating in some way to digital interactivity and design with only very limited text indicating a title, contributor, genre or type of connection, and caption. The core idea of the "Visual Thinking Gallery" is that the photograph—and its quality as a photograph in and of itself—is an important form of knowledge articulation, more important than the text for this form of contribution. A workshop on visual thinking was held at CHI 2012 (Blevis et al., 2012*). In 2012, a primarily visual, image-oriented paper was accepted into the technical program at NordiChi—the first accepted archival paper in a SIGCHI technical program to foreground images over text in articulating its core contribution (Jarvis et al.,* 2012*). And, at DIS 2014, a new pictorials track (Cameron et al.,* 2014*) was introduced in which submitted pictorial essays were reviewed according to standards similar to those of other papers in the technical program; a number of them appeared as archival work. The pictorials submitted to DIS 2014 included a treasure trove of images worth sharing. Blevis et al., 2014 presents a curated selection of the images that appeared in the submitted pictorials with some brief textual descriptions—just enough to state the importance of each image, but not so much as to preclude the images from speaking for themselves. The images combine to form a collection showing a range of different ways in which images serve as first-class exemplars of interaction design, including but not limited to:*

- *images as a record of making*
- *images as a form of making*
- *images as a record of process*
- *images as design ethnography*
- *images as commentary on interaction design*
- *images as purely aesthetic reflection on interactivity*
- *images as social commentary in the perspective of interaction design*
- *images as a record of inspiration*
- *images for reflection to inform a process*
- *images as a record of contexts and environments*
- *images as a record of concepts*
- *images as a record of aesthetic property (i.e., materiality)"*

Now a decade later, many pictorials have appeared which use imagery in ways that include and extend beyond this list. In particular, the rise of generative AI has changed the landscape in ways we consider and contextualize below. In the next section, *Means of Production: By Eye, By Hand, By Prompt, By Speculation*, we speculate on a simple framework based on the work in this volume.

1.4.2 In the Experience of Co-Editor Eli Blevis

In conversations with the editors of the series in which this volume appears, co-editor Eli Blevis recalled a story from the time he was an assistant professor (2002–2007) learning about how to contribute to HCI as someone newly arrived in the area with a background in Design as an instructor and in Computer Science as a doctoral student. He was surprised by the advice he received about publishing in HCI—the advice was not to include too many images. Since that time, he has joined others who—believing in the confluence of HCI with Design—emphasize the role of images and illustrations in the HCI canon. The idea has been to constructively demonstrate the value of images in HCI scholarship, by including in abundance and even privileging images and illustrations over texts.

1.4.3 In the Experience of Co-Editor Miriam Sturdee

Co-editor Miriam Sturdee views pictorials as a way of scaffolding imagination, of enhancing (or replacing!) text, and telling stories about research that cannot be told via the written word alone. The pictorial is not only the visual narrative however, it also better tells physical stories, bringing tangible interaction to the fore.

When Sturdee started in academia, many journals and conferences inactively penalized the inclusion of imagery, in a world of textual expression. She notes that one journal suggested that for every image, you should count it as 250 words (inaccurate as we know a picture is worth a thousand) rather than see it as a way of communicating richness. Within the conference format, for a long time we languished in the idea of page count, rather than word count, and many still hold authors to that —meaning even more images go unseen, and often with complex topics or designerly enquiry, imagery is vital to link the reader to the data.

Communicating such visual stories is also an Art form, and a skill. Yet it is hard to adequately define what makes a pictorial 'successful', and what skills are needed to unpick and review each submission. Hence the variety of our peer reviewed pictorials varies—and it should—inasmuch as our textual offerings do. The way in which we use language and carefully place together words is more practiced by many in interaction design, but we have now the opportunity to share our visual knowledge, in creation and dissemination, to inspire a whole new wave of research and research result.

1.5 Means of Production: *By Eye, By Hand, By Prompt, By Speculation*

There are many ways that pictorials might be classified. Some known. Some, yet to be discovered. There is no single correct way. Notwithstanding, we need a way to contextualize all that is presented in this volume. We introduce the notions of *By Eye*, *By Hand*, *By Prompt*, and *By Speculation* to serve the purpose. The content and approaches to pictorials in this volume are wide and varied. To give a unified account, we focus on the techniques of production that these pictorials comprise.

1.5.1 By Eye

Some of the pictorials in this volume use photographic production techniques. These include photographs as observations, photographs as Art about interaction design, photographs as street photography, photographs as documentation of design and prototyping processes. *By Eye* refers to the means of production which we may think of here as anything involving a light sensitive sensor and a lens and perhaps a viewfinder at capture time, regardless of how post-production is applied. *By Eye* is fundamentally editorial.

A light sensitive sensor may for example be traditional film or a digital sensor. A lens for example may be as primitive as a pin hole or as sophisticated as a modern assembly of glass elements arranged in groups to collect light and correct for aberrations.

Post-production complicates this label. With modern AI-based image production software, such as Adobe Lightroom and Photoshop and in-camera post capture processing, the production techniques are not entirely *By Eye* but also include aspects of *By Hand* in the use of editing tools and filters and *By Prompt* in the use of masking and upscaling and other manipulations of an image that make it not as captured by a lens and sensor.

1.5.2 By Hand

Some of the pictorials in this volume use sketching or physical form production techniques. These include watercolors, pen and ink, collage as recordings of observed experiences, and many types of prototypes as look and feel design and form-giving. *By Hand* refers to the means of production which we may think of here as anything involving drawing or construction. *By Hand* is fundamentally expressive.

Like *By Eye*, *By Hand* is also not entirely clearly defined by boundaries. The wide variety of sketching and prototyping materials complicates this label. For example, a mixing of photographs and sketches in a collage transcends the boundaries between these labels. There are several examples in this volume.

1.5.3 By Prompt

Some of the pictorials in this volume use prompts as production techniques. In this volume, these include generative AI images—*promptographs*, prompts to study participants to produce sketches or physical form prototypes, prompts to collective projects, and prompts to conversational sketching between Artists.

Nowadays, it is tempting to believe that prompt-based AI generative tools are something entirely new. Notwithstanding, prompting has been a part of Design for a long time. *Design* is in the production of a plan or explanation, rather than in the making of something. Whitney (2015):64–65 writes.

> *Moholy-Nagy's approach to these works* [of Art] *differed radically from the typical artistic process of the twentieth century in which a single person conceives, sketches, and makes the artifact in a single stream of activities leading from one to the next. In earlier centuries, when art was among the artisan craft guild traditions, some masters created sketches or engaged in some tasks related to an artwork while assigning other tasks to apprentices. This was also the case in studios, where the same master might undertake works of art, useful artifacts, architecture, jewelry, and other projects. Moholy-Nagy was probably the first artist to give a verbal description to workers who used industrial processes to translate and execute an artwork, when the workers themselves did not aspire to* [become] *artists or artisans. Moholy-Nagy had unbundled the activities of conception and specification from the activity of fabrication.*
>
> *Taking a process that once seemed a single continuous flow, breaking it into steps, and using the steps in new ways is one of the issues that Moholy and other Bauhaus masters explored. In creating these two enamel paintings, Moholy seemed to ask a question: 'If mass production overtakes hand-fabrication, how does this change the nature of conception and specification?' This question was core to the emergence of design as a field distinct from art and craft*

By Prompt is fundamentally Design practice, no matter if the actual implementation of a design is achieved by other people or by machines.

1.5.4 By Speculation

Another means of production that is common in pictorials is speculation about the future that usually takes the form of design fictions (Tanenbaum, Tanenbaum, & Wakkary, 2012), or other forms of *critical design*—that is, *design that makes you think*. In the current volume, this means of production is not as common, but still present. The most salient example of this way of authoring a pictorial is in the chapter by Paulina Yurman, titled *Fluid Speculations*. We provide more detail about this in the section of this introduction chapter that follows, titled *Visual Forms of Knowing and Knowledge.*

1.5.5 Multivalence

By Eye, *By Hand*, *By Prompt*, and *By Speculation* are not mutually exclusive categories. A pictorial chapter can be at once any combination of these. Notwithstanding, a pictorial chapter may more represent one category than another. In the Table 1.1: *By Eye, By Hand, By Prompt, By Speculation,* each pictorial chapter is classified in terms of its production techniques. Large dots are used to denote that one of these categories is most important. Small dots are used to denote that one of these categories figures into the production techniques for that chapter. In the description of each chapter that follows this table, we use shaded icons as well to denote the degree to which each one of these categories applies.

Table 1.1 By eye, by hand, by prompt, by speculation

Production techniques	By Eye	By Hand	By Prompt	By Speculation
Sustainability				
Selfie Time, Wisdom of Years, Dark Hair, and Visible Mending	•			
Seeing Robots Wearing Clothes	•	●	●	•
Conversational Composites: Unexpected Layers	●	•	●	
Seeing and Noticing				
All the Tea in China, Seeing What is and What Can be, Visual Thinking Gallery, Being Photo Visual	•			
Noticing the Rhythms: Women Street Vendor Regulation and Resistance in the Historical Center of Guatemala City	•			
Film in More-Than-Human Design: Using Film to Envision Beyond the Human	•	•		
"You should get to know each other": 10 Years of Sketching in HCI	●	•	●	
Pushing Boundaries				
Design for Respect, Qualities of Focus, Stillness and Motion	•			
Fabric, Beads, and Yarn: Crafting (for) Women	•	•	●	
Fluid Speculations		•		•
Inspirations				
Meaning and Form, Gallery	•			
Plurality, Revisited	•	•	•	
Looking Over the Details	•	•		

(continued)

Table 1.1 (continued)

Production techniques	By Eye	By Hand	By Prompt	By Speculation
For Educators				
For Educators: Design Challenge Based Learning, Quilts, Promptosketches and Promptotypes	•	•	•	
Examining AI-Generated Art in the Educational Context			•	
Sketching Novel Visualization Interaction Metaphors		•		

1.6 Visual Forms of Knowing and Knowledge

1.6.1 Everything is an Object of Knowledge

Nowadays, with the advancement of AI, everything is an object of knowledge. By *everything is an object of knowledge*, we mean that Large Language Model (LLM) AI systems index and contextualize nearly everything that is somehow represented online and make nearly everything available through search and prompts. Such systems reduce the need for anyone to experience anything on their own with their own eyes or by their own hand. By *everything*, we mean *everything*—not just texts, but also images, and other forms of representation of things real or imagined.

1.6.2 Knowing the Difference Between What is Real and What is Imaginary is Harder Nowadays

When knowledge of everything—real or imagined—can be translated *By Prompt* into anything—real or imagined, by means of generative AI, the very idea of knowledge creation is uncertain. Moreover, we cannot easily know the difference between what is real and what is imagined, anymore.

1.6.3 Authenticity is the Currency of Human Knowledge: Acts of Seeing Are Acts of Knowing

Nowadays, the currency of humanity is authenticity. Authenticity alone distinguishes between human knowledge and machine learning. Authenticity means that we have actually experienced something firsthand, that we *know* about it firsthand. In this book, the authors of the individual chapters are mostly expressing accounts of their firsthand experiences, *By Eye* or *By Hand*. That is to say, they are translating from their experiences into visual language and the act of this translation, we argue, is an act of knowledge creation. The authors are *seeing* interaction design and each act of seeing is an act of knowing.

1.6.4 Acts of Knowing Everything at Hand

This framing—each act of seeing is an act of knowing–corresponds to an intuitionistic perspective in the area of formal logics—a branch of Mathematics and a foundation of Computer Science. Intuitionism is also known as constructivism (as a term of Art in mathematics, rather than its use as a term in Education), and it means reasoning with what is at hand. The mathematician Per Martin-Löf (1987) explains that a proof is "an act of knowing an object of knowledge." A proof is valid if it is evidenced by truth. In Design, this is similar to Donald Schön's notions of Design as *bricolage*—meaning that Designers work with materials at hand [Schön (1979) as quoted in Nelson and Stolterman (2014)]. This line of argument has similarities with Barthes (1977) accounts of how images, music, texts create meanings. We could argue that LLMs and generative AI are also constructivist—that is they work by learning from everything that is at hand, albeit at super-human scale and albeit regardless of if the digital materials they learn from represent real or imagined things. We write more about this below.

1.6.5 Real Things and Imaginary Things

There are some aspects of pictorials that are not *By Eye* or *By Hand*. Here, we write that some are *By Prompt* or *By Speculation. By Prompt* includes the notion that rather than relying on one's own experiences, we can query an LLM using generative AI to produce forms—images or texts or sounds, for example—that may or may not owe in any way to real, observable things. *By Speculation* includes the notion at best that we predict futures by means of constructing forms—again, images or texts or sounds, for example—in order to imagine *alternative futures* [the term owes to Fry (1999)]. In Design, speculation includes the notion of *speculative everything* which owes to the work of Dunne and Raby (2024, recently revised). *By Speculation* also includes the notion of *design fictions* which owe to Tanenbaum, Tanenbaum, & Wakkary (2012).

The use of design fictions in pictorials is described in Lewis and Sturdee (2024). *By Eye* or *By Hand* can also produce speculative forms in addition to representations of actual experiences and knowledge. What is important is that a person producing forms *By Eye* or *By Hand* knows if such forms represent real or imaginary things. A person producing forms *By Prompt* may not really know for sure if such forms represent real or imaginary things. A person producing forms *By Speculation* knows that such forms are not—at least not yet—real.

In a classical sense, an image or text or another form represents knowledge just in the case that it follows from observable things or experiences within our own cognition and perceptions (after Martin-Löf, 1987). In a world in which many forms owe to prompts or speculations, we are best advised to be cautious about truth and knowledge. That is, don't trust everything you hear, everything you read, and sometimes you can't even trust things you see.

1.6.6 Ambiguity and Proliferation of Meanings

When people—scholars of the social sciences in particular—speak about knowledge creation, they do not generally mean to include the objects of knowledge, but rather the understanding of meanings that objects of knowledge entail, more often than not expressed in language. In AI systems, we do not really need to distinguish between objects of knowledge and their meanings in this way, anymore. These systems freely translate between one form and another. A textual prompt is easily turned into an image. An image is easily described in text. Are such translations entirely accurate? Of course, they are not. There are many ambiguities introduced when translating from one syntactic form to another. Is a textual description less ambiguous than a visual form? Some will always argue that textual descriptions are knowledge in ways that are less ambiguous than visual forms. Notwithstanding, others may hold that a visual form may be knowledge in and of itself. We can also argue that both textual descriptions and visual forms are very ambiguous, and far more so than formal mathematical languages, which alone provide for unambiguous understanding. Sengers and Gaver (2006) argue that multiple meanings are inherent to HCI and Design and may be understood to be a strength rather than a weakness.

1.7 A Few Things, Finally

1.7.1 The Role of the Visual in HCI and Interaction Design

Pictorials are an independent form, distinct from purely text-based scholarship. *By Eye, By Hand, By Prompt,* and *By Speculation* are methods of production that may apply to any Design-oriented research, scholarship, or creative activity. They do not replace text, but they are not or best not illustrations nor decorations of text either.

The acceptance of pictorial forms in HCI and Interaction Design scholarly venues is now well established. As in any academic area, there are some who may argue that only text is unambiguous scholarship. In this volume, we take the point of view that visual materials are objects of knowledge in their own right. Our reasoning here is constructive in the sense that in making a volume that emphasizes visual materials over texts simply makes the importance of visual materials apparent. Now, it is up to the viewer/reader to decide.

1.7.2 Alternative Forms of Classification

In terms of a taxonomy of pictorials, there are several exogenic surveys of the archive of pictorials to date. Like nearly all surveys, these can be reductive. In some ways, such surveys do a disservice especially in the case of pictorials wherein the *ultimate particularity* of the images and illustrations and choice of how little or how much text accompaniment are part of an experience that is somewhere between scholarship and practice. The term *ultimate particularity* as a notion of Design's primary concern owes to Nelson and Stolterman (2014). In some other ways, these surveys are very welcome as a means of welcoming others to the form. Among the surveys of pictorials that have appeared, Yin et al. (2024) identifies the myriad accessibility issues in many pictorials. Kirkegaard Rasmussen et al. (2024) is the most recent survey we know of, and it counts 271 pictorials in its corpus of analysis. This analysis posits seven analytical themes: *"(1) composition, (2) typography, (3) colour, (4) drawing attention, (5) visual style, (6) active engagement, and (7) visual related work."*

1.7.3 Text and Images

The seminal treatise on the difference and interplay between Texts and Images is Barthes' "Image, Music, Text" (Barthes, 1977). Barthes' semiotic approach is not light reading for many in HCI or Design. In computer science and mathematics, we can point to Martin-Löf (1987) for an account of the semantics of formal systems as model-theoretic translations between syntactic systems. Herein, we may more generally understand visual materials and texts to some extent as syntactic systems, in addition to mathematics. Martin-Löf's approach is also not easy reading for many in HCI and Design.

The problem of translating from text to images and images to text is also a matter of accessibility as in Yin et al.'s (2024) pictorial we cite above. In HCI, Alt-tags associated with non-textual elements of scholarship and interaction design artefacts alike are a central concern of the discipline. Alt-tags are hard to write for someone concerned with making visual materials accessible and doing so is not easily automated, even with advances in AI. Moreover, alt-tags are oftentimes hard to interpret for someone blind or low-sighted.

Koskinen (2004) also takes an approach to text and images that resonates with HCI and Interaction Design. Koskinen describes the transitions that have taken place in the use of mobile devices at the time for image making as an everyday, *naturalistic* catalyst to conversations between ordinary people. Koskinen uses the term *perpetually* to denote the everywhere, every-day, everything roles emergent in an early study of mobile interactions with images. Mobile interactions with images have become ubiquitous in our interactive lives, nowadays.

In the pictorials in this volume, the texts serve to describe the visual material. Visual materials are herein privileged. In this volume, the best we hope for is that the textual descriptions and annotations of the visual materials are sound—that is, they owe to our semantic interpretations of the world. We cannot hope that they are complete in the sense that they account for all reasonable meanings.

1.8 Pictorials: The First Decade

Restating by way of summary, *pictorial* is the term developed in HCI over the past decade to denote a visual essay or paper in which the visual material is at least co-equal in importance to the text. The origins of the term are described in Blevis et al. (2015). The first pictorial published in HCI is Jarvis et al. (2012).

Is *pictorial* the right word for such a visual essay? Not necessarily. The term *pictorial* is possibly too broad. It is possibly too informal or too readily associated with illustration. It may refer to work that has scholarly merit. It may refer to work that does not have scholarly merit. The notion of *scholarly merit* itself is not easily defined and varies widely from one area of scholarship to another.

We may prefer the term *visual essay*. We may prefer the term *visual essay* because the ideals of a good essay imply good scholarship. An essay is not necessarily scientific reporting. An essay can and should make use of scientific reporting. An essay argues *for* something. Visual essays argue *for* something.

Regardless of these preferences, *pictorial* is the word that has come to be used in Human–Computer Interaction (HCI) in which the work in this volume is primarily situated. Some of the pictorials that appear in HCI currently use visual materials as illustrations. All of the chapters in this volume and many pictorials in HCI treat visual materials as first-class content in and of themselves.

1.9 Synopses

In the remainder of this chapter, we briefly discuss each of the chapters with a selected thumbnail image from each one to characterize the style of each one.

1.10 Sustainability

Selfie Time, Wisdom of Years, Dark Hair, and Visible Mending

Shunying an Blevis (Indiana University, US) and Eli Blevis (Indiana University, US)

In the first chapter of the sustainability section, namely *Selfie Time, Wisdom of Years, Dark Hair, and Visible Mending,* Blevis and Blevis present a curation of photographs selected from various prior pictorials engaging with notions of sustainability. The collection is tuned to present points in the affine space of *seeing* sustainability in photographs. The approach is to make a few visual points in some depth, rather than a holistic treatment of the entire possibilities of understanding sustainability in photographs. The authors present a mix of posed, staged photographs, opportunistic photographs, and photographs representing very different scales in perspectives from the Hong Kong skyline to the threads in a mended pair of jeans, for example. The thumbnail we have chosen for this introductory chapter shows a petrol station in the Hung Hom district of Hong Kong with a curious translation on the sign towards the middle of the frame, "No Naked Lights." (Thumbnail 1.1)

By Eye *By Hand* *By Prompt* *By Speculation*

•

Thumbnail 1.1 Sustainability. Blevis & Blevis, *Selfie Time, Wisdom of Years, Dark Hair, and Visible Mending*

Seeing Robots Wearing Clothes
Zaiqiao Ye (Harbin Institute of Technology, CN)

In *Seeing Robots Wearing Clothes*, Zaiqiao Ye pictures a future in which fashion for robots is an opportunity for advancing notions of fashion that are compatible with sustainable futures rather than forces in opposition. Ye takes us on a journey through fashion accessories for Aibo robots. In one image, she highlights the keyhole design of one outfit for Aibo that functions as an opening for its camera. Ye reports on a study in which she asks participants to play a game relating to fashion for robots as a means of uncovering attitudes towards robot fashion, especially with respect to sustainability. We've selected Thumbnail 1.2 to show the sketches that her study participants created. The thumbnail image appears in her chapter with the title "Participants' Imagination of Their Personal Robots' Clothes." The chapter has a mix of mainly photographs with some sketches by and prompts for study participants who are themselves engaged in speculative futures (Thumbnail 1.2).

Thumbnail 1.2 Sustainability. Ye, *Seeing Robots Wearing Clothes*

Conversational Composites: Unexpected Layers

Mafalda Gamboa (Chalmers University of Technology, SE), Sara Ljungblad (Chalmers University of Technology, SE), and Miriam Sturdee (University of St. Andrews, UK)

The authors state, "We present the story of Conversational Composites: a flexible method grounded in the material and tangible qualities of sketching in different forms and media, creating physical and digital layers of conversation between participants." The primary production techniques modality is sketching, with prompts and photographs also playing a role (Thumbnail 1.3).

Thumbnail 1.3 Sustainability. Gamboa, Ljungblad, & Sturdee, *Conversational Composites: Unexpected Layers*

1.11 Seeing and Noticing

All the Tea in China, Seeing What is and What Can Be, Visual Thinking Gallery, Being Photo Visual Shunying an Blevis (Indiana University, US) and Eli Blevis (Indiana University, US)

This chapter is a tour of mostly opportunistic street photography appearing in prior conference pictorial publications, wherein street photography is understood as Artwork. All but one of the images are street photography. One is staged, also as a form of photographic Artwork. The approach may be distinguished from visual ethnography because (a) it lacks the systematicity of the social sciences approach and (b) the quality of the photographs as Artwork matters. Notwithstanding, each photograph in the chapter is taken up as an opportunity to see details that we might otherwise miss in the day to day of everyday. Blevis and Blevis also describe their practice of annotating photographs, a practice very much inspired by notions of annotation introduced into the HCI archive by William Gaver and his colleagues (Thumbnail 1.4).

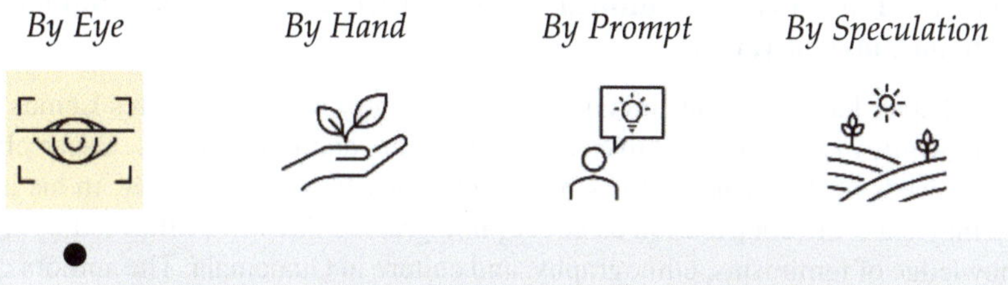

Thumbnail 1.4 Seeing and Noticing. Blevis & Blevis, *All the Tea in China, Seeing What is and What Can be, Visual Thinking Gallery, Being Photo Visual*

Noticing the Rhythms: Women Street Vendor Regulation and Resistance in the Historical Center of Guatemala City

Gustavo Fernandez (Guatemalan Photographer and Researcher, GT), Andrea Echeverria (Guatemalan Researcher, GT), Katreen Boustani (Syracuse University, US), Lucy Lemus (Cultural Advisor, US), and Christena Nippert-Eng (Indiana University, US)

This chapter has been contributed in loving memory of Oscar Andrés Lemus who, in the words of the authors, "lives among the stars and in our hearts. Oscar planted all the seeds of this project; it has been a great joy to help them grow in his honor. We may class the contribution as street photographic Artwork, informed by a deep knowledge of feminisms, ethnography, and culture in Guatemala. The authors create an account of the daily life of street vendors in Guatemala City's *Centro Histórico.* The photographs take up an historically informed perspective on the center's colonial and post-colonial forms and oftentimes tragic histories as embodied in, for example, the account of the name, *Plaza de las Niñas*. The images taken over nine months from many different angles are a conversation that centers street vendors in relation to rights of place among all of the historical and political features of this place (Thumbnail 1.5).

Thumbnail 1.5 Seeing and Noticing. Boustani, Echeverria, Fernandez, Lemus, & Nippert-Eng, *El Turno Malo: Noticing the day-to-day rhythm of street vendor regulation and resistance in the historical Center of Guatemala City*

Film in More-Than-Human Design: Using Film to Envision Beyond the Human

Heidi Biggs (Georgia Institute of Technology, US), Doenja Oogjes (Eindhoven Institute of Technology, NL), and Shaowen Bardzell (Georgia Institute of Technology, US)

Heidi Biggs, Doenja Oogjes, and Shaowen Bardzell take up notions of decentering the human in design to develop a posthuman gaze. The posthuman gaze helps us understand the interdependencies or entanglements between everything and everything else, as an alternative to "user and used." The importance of the more-than-human ontological perspective is made all the more salient by, for example, climate change and the rise of AI. Inspired by film theory, especially owing to Gilles Deleuze, the authors explore everyday video using smartphones as a method to investigate their more-than-human acts of noticing. The many images in this chapter combine sketches of how our videographer, researcher authors have set up their smartphone cameras with stills from the videos produced. These are proposed as a method for understanding and sharing their more-than-human gaze (Thumbnail 1.6).

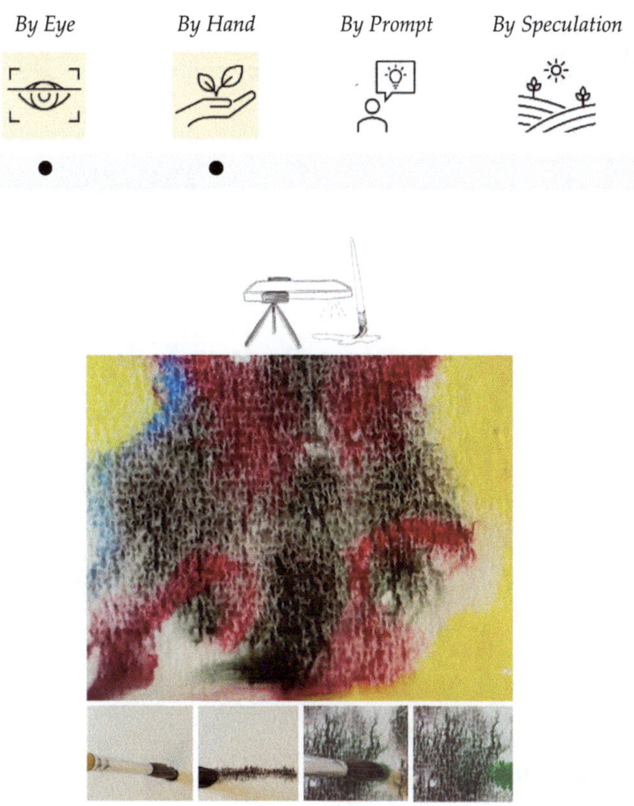

Thumbnail 1.6 Seeing and noticing. Biggs, Ologies, & Bardzell, *film in more-than-human design: using film to envision beyond human in design research*

"You Should Get to Know Each Other": 10 Years of Sketching in HCI.

Makayla Lewis (Kingston University, UK) and Miriam Sturdee (University of St. Andrews, UK)

This is a chapter about friendship forged from a common interest in sketching as a means of seeing interaction design. Sturdee and Lewis present a retrospective on their decade of intersecting practice, research, and education. For each image or set of images, they present a takeaway message concerning collaboration, creativity, mutual support, and the ongoing nature of possibilities that emerge from collaborating in this way. Image 8 titled "Spending a lot, or too much, time together" is a particularly compelling record of this friendship. As can be seen elsewhere in this book, Sturdee and Lewis have also led the way in fostering collectives that engage in bringing Art to HCI and interaction design. The production techniques are primarily sketching *By Hand* with some use of photographs as a material of collage (Thumbnail 1.7).

Thumbnail 1.7 Seeing and noticing. Miriam Sturdee & Makayla Lewis, *Sketching (HCI)*

1.12 Pushing Boundaries

Design for Respect, Qualities of Focus, Stillness and Motion

Shunying an Blevis (Indiana University, US) and Eli Blevis (Indiana University, US)

The images in this chapter take up themes at the boundaries of what may be considered to be interaction design in ways that entail political interpretations. Blevis and Blevis offer photographs of protest signage. They offer satiric photographic design fictions that appeal to notions of design for respect. There are also extremely defocused images that push at the boundaries of photographic truth and cognition. Finally, there are images from Blevis (2014), a pictorial that was among the first Association for Computing Machinery (ACM) pictorials track which appeared at the Design Interactive Systems (DIS) conference in that year. The images compare consumerism in general with consumerism in the digital world, ending with a commentary on sustainability (Thumbnail 1.8).

Thumbnail 1.8 Pushing Boundaries. Blevis & Blevis, design for respect, qualities of focus, stillness and motion

Fabric, Beads, and Yarn: Crafting (for) Women's Orgasm

Dianya Mia Hua (the Pennsylvania State University, US)

In this chapter, Dianya Mia Hua reports on a participatory design study of artefacts created for women's pleasure by women who participated in this study. The goal is very much an idea for celebrating and reclaiming women's autonomy and power. Hua describes the study in relation to feminism history. Her work is activist and rightly so—Hua explains that the personal is also political. Hua's reporting recounts the discussion of the participants in a highly engaging way. These discussions range from describing the artefacts as cute to artefacts with agency to reconnect with one's body after disturbing non-consensual sexual encounters. The images themselves are quite beautiful, even if one does not immediately recognize their function. Important work. The production techniques are photographs of artifacts created *By Hand* by participants (Thumbnail 1.9).

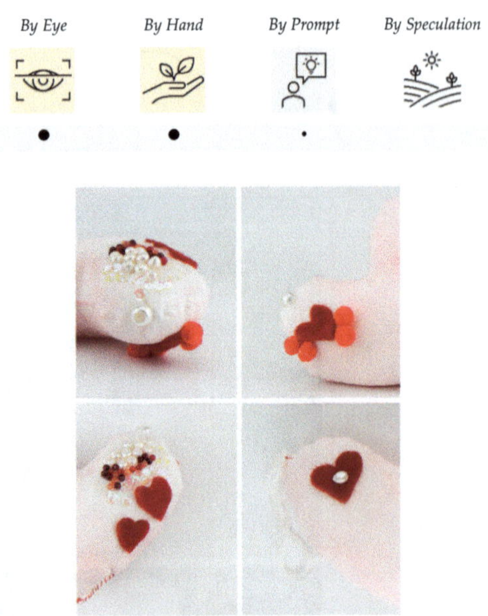

Thumbnail 1.9 Pushing Boundaries. Hua, Fabric, Beads, and Yarn: Crafting for Women's Orgasm

Fluid Speculations

Paulina Yurman (University of the Arts London, UK)

Paulina Yurman contributes a chapter of truly compelling watercolor drawings, which in her words are "experiments in drawing artefacts and technologies." Yurman points out that there are already rich traditions of scholarship that argue for "drawing as a form of discovery and knowledge production." She shares an account of the history of these contributions. As a way of seeing, Yurman argues that the imprecision of watercolors is a feature of their expressive power in the context of sketching design artefacts. Moreover, Yurman states "the drawings presented here show how artefacts and technologies have been looked at through a process of drawing that merges knowledge about them as they currently exist with ideations about how they might also be understood (or misunderstood)." The drawings here take up notions of narrative and free association, adorning machines, speculation about bodily fluids, and machine interpretations of drawings. Yurman also provides a feminist analysis of many of these themes and drawings. Finally, we can share here that this is award winning work in the newly developing world of pictorials that have appeared in HCI's archival venues. The production techniques are sketches of the Artist's observations about mostly ordinary and local artifacts, which the Artist herself describes as speculations (Thumbnail 1.10).

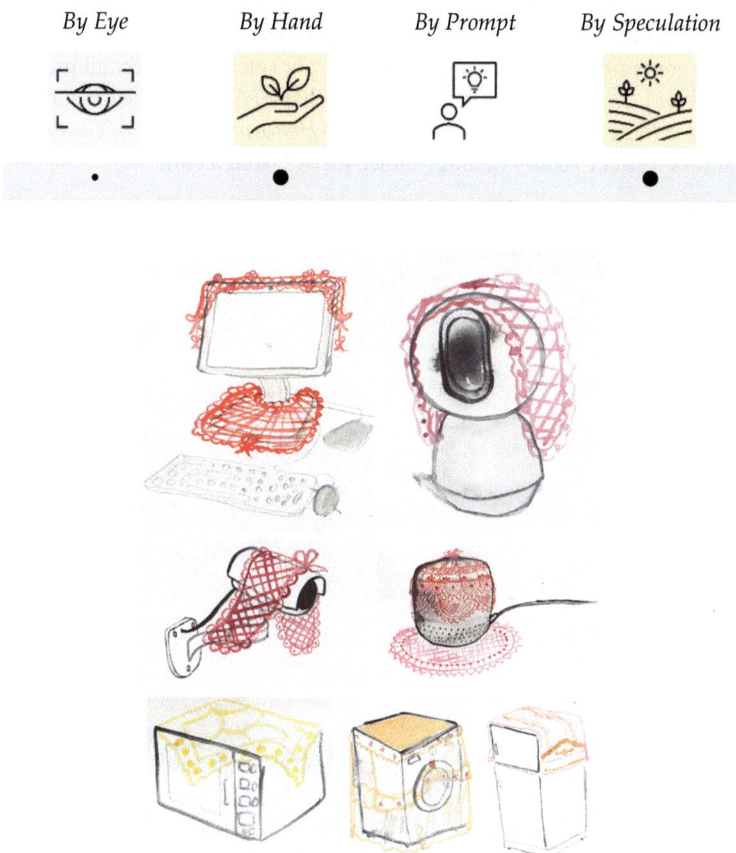

Thumbnail 1.10 Pushing Boundaries. Yurman, *Fluid Speculations*

1.13 Inspirations

Meaning and Form, Gallery

Shunying an Blevis (Indiana University, US) and Eli Blevis (Indiana University, US)

This chapter shares mostly early work by Blevis and Blevis. These images are among the first ones they were able to publish in interaction design venues, some appearing as early as 2011. Many appeared in the Visual Thinking Backpage Gallery which appeared in each issues of the ACM interactions periodical beginning in 2011 and continuing under new curatorship into the present. Some are new material. The role of these images in relation to interaction design is to scaffold a thesis about the importance of quality in image making and a thesis that photographs in addition to their Artistic value may be objects of knowledge in and of themselves. The production techniques are entirely photographic. Compare our selected thumbnail image below, Thumbnail 1.11*: Inspirations*, to image 6 in the chapter *Noticing the Rhythms: Women Street Vendor Regulation and Resistance in the Historical Center of Guatemala City* which shows a street vendor selling "elotes locos" from a cart.

Thumbnail 1.11
Inspirations. Blevis & Blevis, *Meaning and Form, Gallery*

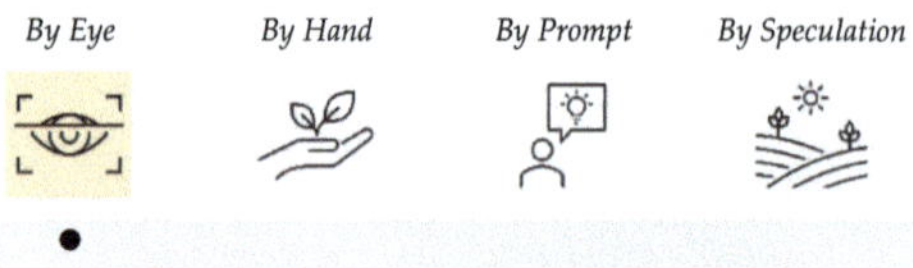

Plurality, Revisited

Miriam Sturdee (University of St. Andrews, UK), Makayla Lewis (Kingston University, UK), Josh Urban Davis (Dartmouth College, US), Angelika Strohmayer (Northumbria University, UK), Sarah Fdili Alaoui (University of

the Arts London, UK & Université Paris Saclay, FR), Nantia Koulidou (Sheffield Hallam University, UK), Katta Spiel (Technical University of Wien, AT).

This chapter by a collective of authors asks a fundamental question about the nature and interaction of Art and research. It is a personal account of the various approaches within the collective take to being both Artists and researchers. It is a further reflection on having asked this question before and the merits of asking again. Rather than advancing a single answer, the authors invite us to consider a plurality of artistic and research practices in this reflection on four years of work.

Miriam Sturdee reflects on various ways to reclaim time for her expression, while still under the obligations of a scholarly researcher. She contributes, for example, a beautiful watercolor of old vines in a Japanese style she did from a photograph she had taken a year earlier. She shares her joy in finding the time to make this transformation from photograph to Artwork.

Makaya Lewis shares her experience creating sketch prompts for others over years, and a newfound embrace of celebrating her own sketches as a form of autoethnography. Indeed, Dr. Lewis has created many opportunities for others as a very generous act for which heartfelt thanks are in order. We hold that her focus—together with Miriam Sturdee and others—on fostering collective work is an innovation in seeing interaction design.

Josh Urban Davis articulates a personal experience of both combining and creating "breathing room" between their Art and research practices. Dr. Davis's tarot card images illustrate technological and Artistic choices. About these, they write "In both failed utopian visions, the one promised by traditional faiths and the one promised by emerging technologies, something was lost along the way. A choice was made. A world was deleted. The Deleted World Tarot is a tool for healing the rift between the choices of the past and the consequential loss of alternative futures. What is not saved will be lost."

Angelika Strohmayer's threads and beads on canvas denote the chaos she sometimes feels towards her reflection on being both an Artist and researcher. This chaos also unifies her understanding of her experiences, writing "like this chaos. I like the mess. Both my hoop and my written documents need to be fixed. But that's where my thoughts seem to thrive—the mess of research, the mess of funding applications, the mess of unplanned stitches."

Sarah Fdili Alaoui is a performance and dance Artist. Vintage, a piece of her own choreography is illustrated here. She emphasizes the de-centering of technology.

Nantia Koulidou records her work on material explorations and experiences. These take the form of digital jewelry. She discusses the difficulties in having her reporting on these forms accepted into HCI archives—a difficulty all of the authors in this book share and work towards overcoming.

Finally, Katta Spiel contributes beautifully executed line drawings on colored backgrounds of disability cultures in interaction with technologies. Her images are produced on a tablet and focus on specific individuals, rather than abstract concepts (Thumbnail 1.12).

This chapter illustrates production techniques, *By Eye*, *By Hand*, and *By Prompt*.

Thumbnail 1.12 Inspirations. Sturdee, Lewis, Davis, Strohmayer, Alaoui, Koulidou, & Spiel, *Plurality, Revisited*

Looking Over the Details

Andy Boucher (Northumbria University, UK)

The very idea of pictorials in the world of HCI and interaction design has at its foundation the first pictorial essay to make it into the ACM conference archive—a paper that pioneered the foregrounding of visual materials titled "Attention to detail: annotations of a design process" (Jarvis, Cameron, & Boucher, 2012). In this chapter, one of these original authors, Andy Boucher, contributes a retrospective on this original contribution. Boucher explains that the original work arose from a need to be able to illustrate their design process in a way that was not able to be accommodated in a (still nowadays) traditional notion of scholarly archival work in HCI. The images presented then and now here are a visual curation of the decisions made during the design and prototyping of an indoor weather station. All of us who have submitted our visually privileged work to such venues have endured some resistance from reviewers and some of it quite insensitive, as Boucher illustrates in the text of this chapter. Nowadays, pictorials are a common form of archival contribution in many ACM conferences, a form of progress we owe to Dr. Boucher and his colleagues.

The images and their descriptions in this chapter are compelling. Boucher offers descriptions of material choices, revisions, sketches, 3d printed prototypes, cultural probes, experiments, and other elements of the design process. The photographs themselves are quite lovely to look at as photographs in and of themselves—a feature of this work that must not be overlooked. Boucher aptly states "The essay, once a novel provocation, has become a historical artifact." We are delighted to be able to include it in this book. The production techniques are primarily photographic documentation of physical, by-hand prototypes (Thumbnail 1.13).

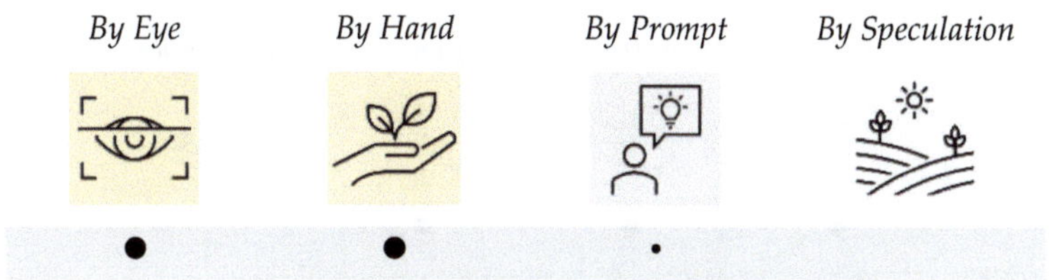

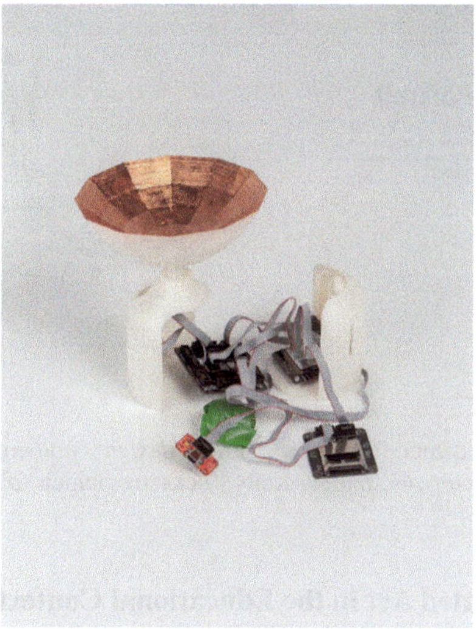

Thumbnail 1.13 Inspirations. Boucher, *Looking Over the Details*

1.14 For Educators

For Educators: Design Challenge Based Learning, Quilts, Promptosketches and Promptotypes

Shunying an Blevis (Indiana University, US) and Eli Blevis (Indiana University, US)

Blevis and Blevis share a small sampling of classwork—some published, and all from students who have already graduated from a program in Human–Computer Interaction Design. This classwork all origins in assignments that emphasize visual literacy in the context of HCI and interaction design. Sharing them here is intended as possible inspirations for others in their teaching. The production techniques includes *By Eye*, *By Hand*, and *By Prompt* (Thumbnail 1.14).

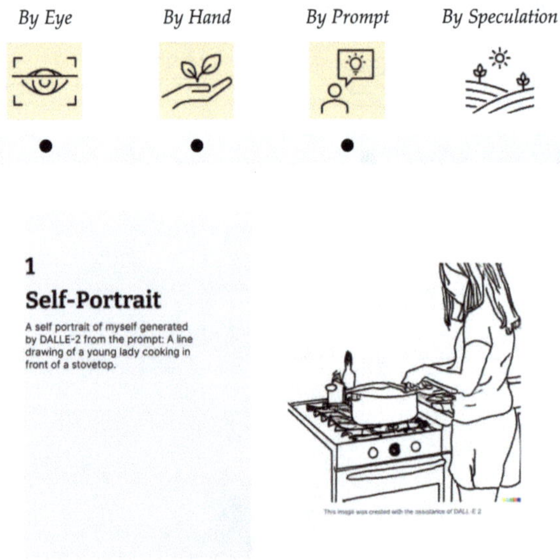

Thumbnail 1.14 For Educators. Blevis & Blevis, *design challenge based learning, quilts, promptosketches and promptotypes. Image:* Kelly Lucksanawichien, 2023, used with permission

Examining AI-Generated Art in the Educational Context

Akesha Horton (Indiana University, US)

Akesha Horton provides a survey of current questions about AI generated Art, specifically for the context of Arts & Crafts educators. A professional crafter herself, Dr. Horton considers the advantages and challenges of these newly trending technologies, most of the references are quite recent. Following a definition by McTighe & Wiggins, Horton poses essential questions about this new technological frontiers' implication for practice. Rather than answer these questions generally, Horton answers them constructively with beautiful AI generated paper quilting from her own Art practice and communities. Her images combine the curves of the Sydney Opera house with floral imaginings. Her constructive approach yields new questions about her results from the Midjourney software she used. The images also show a progression of detail in the prompts she provided for her *promptoquilling*. Horton notes that it takes time and thoughtfulness to create such prompts. This is the only chapter in the present volume which is entirely produced *By Prompt*s (Thumbnail 1.15).

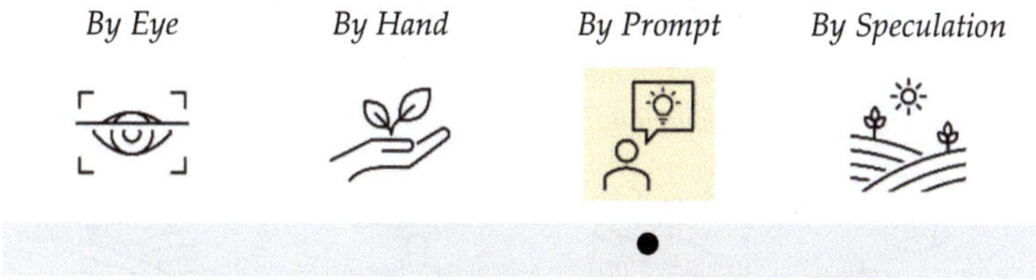

Thumbnail 1.15 For Educators. Horton, Examining *AI-Generated Art in the Educational Context*

Sketching Novel Visualization Interaction Metaphors

Richard Brath (Uncharted, CA), Johanna Drucker (UCLA, US), Christophe Schuwey (Université Bretagne Sud, FR), Joris van Zundert (Huygens Instituut, NL), Johannes Liem (University for Continuing Education Krems, AT).

Brath et al. share drawings from a workshop designed to push boundaries in data visualizations. It is the idea of the workshop that situates this chapter in the "For Educators" section. The chapter could equally have been placed in the "Pushing Boundaries" section. The authors describe the uneasy fit for humanities inspired data visualizations compared with the conventions of data visualizations in the sciences. The result is a delightful collection of data visualization drawings that escape the strictures of conventional data visualization forms, such as network diagrams, bar charts and so forth. The authors argue that when learners start with such conventional forms, their design processes are constrained. They term their provocation targeted at escaping these boundaries as a "wacky encoding exercise." The production techniques are entirely sketching, *By Hand* (Thumbnail 1.16).

Thumbnail 1.16 For Educators. Brath, Drucker, Schuwey, van Zunder, & Liem., *Sketching Novel Visual Interaction Metaphors*

Acknowledgements We especially acknowledge and thank the anonymous reviewers and the series editors, Ilpo Koskinen and Peter Gall Krogh, for many valuable insights and conversations about this project.

References

Barthes, R. (1977). *Image Music Text: Essays Selected and Translated by Stephen Heath*. Fontana Press.

Blevis, E., Hauser, S., & Odom, W. (2015). Sharing the hidden treasure in pictorials. *Interactions, 22*(3), 32–43. https://doi.org/10.1145/2755534

Blevis, E. (2014). Stillness and motion, meaning and form. In *Proceedings of the 2014 Conference on Designing Interactive Systems (DIS '14)* (pp. 493–502). Association for Computing Machinery. https://doi.org/10.1145/2598510.2602963

Blevis, E., Churchill, E., Odom, W., Pierce, J., Roedl, D., & Wakkary, R. (2012). Visual thinking & digital imagery. In *CHI '12 Extended Abstracts on Human Factors in Computing Systems (CHI EA '12)* (pp. 2715–2718). Association for Computing Machinery. https://doi.org/10.1145/2212776.2212703

Blevis, E. (2011). Digital imagery as meaning and form in HCI and design: an introduction to the Visual Thinking Backpage Gallery. *Interactions, 18*(5), 60–65. https://doi.org/10.1145/2008176.2008190

Blevis, E., Lim, Y.-K., Ozakca, M., & Aneja, S. (2005). Designing interactivity for the specific context of designerly collaborations. In *CHI '05 Extended Abstracts on Human Factors in Computing Systems* (*CHI EA '05*) (pp. 1216–1219). Association for Computing Machinery. https://doi.org/10.1145/1056808.1056880

Blythe, M. (2014). Research through design fiction: Narrative in real and imaginary abstracts. In *Proceedings of the SIGCHI Conference on Human Factors in Computing Systems* (pp. 703–712).

Cameron, D., Hauser, S., Jarvis, N., & Odom, W. (Eds.). (2014). Pictorials. In *Proceedings of the 2014 Conference on Designing Interactive Systems* (pp. 121–160 and 473–502). ACM.

Drucker, J. (2014). *Graphesis: Visual Forms of Knowledge Production* (Vol. 6). Harvard University Press.

Dunne, A., & Raby, F. (2024). *Speculative Everything*. Design, Fiction, and Social Dreaming. MIT Press.

Fry, T. (1999). *A New Design Philosophy: An Introduction to Defuturing*. UNSW Press.

Jarvis, N., Cameron, D., & Boucher, A. (2012). Attention to detail: Annotations of a design process. In *Proceedings of the 7th Nordic Conference on Human-Computer Interaction: Making Sense Through Design (NordiCHI '12)* (pp. 11–20). Association for Computing Machinery. https://doi.org/10.1145/2399016.2399019

Koskinen, I. (2004). Seeing with Mobile Images: Towards Perpetual Visual Contact. In *Proceedings of The Global and the Local in Mobile Communication: Places, Images, People, Connections. COMMUNICATIONS IN THE 21ST CENTURY: The Mobile Information Society, Conference*. Budapest, June 10–12, 2004.

Lee, T. B., & Trott, S. (2023). Large Language Models, Explained with a Minimum Of Math and jargon. https://www.understandingai.org/p/large-language-models-explained-with accessed 06.17.2025.

Lewis, M., & Sturdee, M. (2024). Design Fiction and Speculative Sketching. In *Sketching in Human Computer Interaction: A Practical Guide to Sketching Theory and Application* (pp. 125–141). Springer Nature Switzerland.

Martin-Löf, P. (1987). Truth of a proposition, evidence of a judgement, validity of a proof. *Synthase, 1987*, 407–420.

Harold, G. (2014). *Nelson, and Erik Stolterman*. Intentional Change in an Unpredictable World. MIT Press.

Rasmussen, M. K., Pakanen, M., & Heiselberg, K. (2024). A review of visual strategies in pictorials. In *Proceedings of the 16th Conference on Creativity & Cognition (C&C '24)* (pp. 211–228). Association for Computing Machinery. https://doi.org/10.1145/3635636.3656212

Schön, D. A. (2017). (first published 1979). *The Reflective Practitioner: How Professionals Think in Action*. Routledge.

Sengers, P., & Gaver, B. (2006). Staying open to interpretation: engaging multiple meanings in design and evaluation. In *Proceedings of the 6th Conference on Designing Interactive Systems (DIS '06)* (pp. 99–108). Association for Computing Machinery. https://doi.org/10.1145/1142405.1142422

Tanenbaum, T. J., Tanenbaum, K., & Wakkary, R. (2012). Design fictions. In *Proceedings of the Sixth International Conference on Tangible, Embedded and Embodied Interaction* (pp. 347–350).

Whitney, P. (2015). Design and the Economy of Choice. *She Ji: THe Journal of Design, Economics, and Innovation, 1*(1), 58–80.

Yin, A., Sogani, R., Oewel, B., Phan, K., Park, J. S. Y., Yeo, M. A., Yazzolino, L. A., Arcos, K., Abdolrahmani, A., Blank, E., Gilbert, M. D., & Branham, S. M. (2024). "Malicious" Pictorials: How Alt Text Matters to Screen Reader Users' Experience of Image-Dense Media. In *Proceedings of the 2024 ACM Designing Interactive Systems Conference (DIS '24)* (pp. 1262–1274). Association for Computing Machinery. https://doi.org/10.1145/3643834.3660747

Part I
Sustainability

Chapter 2
Selfie Time, Wisdom of Years, Dark Hair, and Visible Mending

Shunying An Blevis and Eli Blevis

Abstract Seeing interaction design as a matter of understanding sustainability is taken up in this chapter. Over years, the authors have contributed many papers—some visual and some more traditional—on the nature and imperatives of understanding sustainability as a critical focus of interaction design (including HCI and Design). In this chapter, we have selected several images from the papers and pictorials listed above. The selection is based on sampling this work but does not replace the original work. Moreover, in including images and descriptions here, we are illustrating selected aspects of sustainability that are visually compelling. We are in no way offering a complete account of sustainability in interaction design.

2.1 Contributions

There are many ways of picturing sustainability in interaction design. Herein, we provide a few interesting vignettes in this space concerning topics of spending too much time engaged with personal technologies, making mending visible as a matter of sustainability, foraging, and signage.

Image 2.1: Only Tablet with Dark Hair (A), Image 2.2: Only Tablet with Dark Hair (B), Image 2.3: Only Tablet with Dark Hair (C).

In *Only Tablet with Dark Hair* (Blevis & Blevis, 2022a), we take up the unsustainability of technology dependence in the realm of contemporary individual lifestyles.

Photographs in this chapter are selected from *Selfie Time* (Blevis, 2018), *Wisdom of Years* (Blevis and Blevis, 2018), *Only Tablet with Dark Hair* (Blevis & Blevis, 2022a), *and Visible Mending* (Blevis & Blevis, 2022b).

S. A. Blevis · E. Blevis (✉)
Luddy School of Informatics, Computing, and Engineering, Indiana University at Bloomington, Bloomington, USA
e-mail: eblevis@iu.edu

S. A. Blevis
e-mail: san@iu.edu

S. An Blevis et al. (eds.), *Seeing Interaction Design*, Design Research Foundations,
https://doi.org/10.1007/978-3-032-14200-9_2

Image 2.1 Only Tablet with Dark Hair (A), Shunying An Blevis and Eli Blevis, Photograph, 2020

Image 2.2 Only Tablet with Dark Hair (B), Shunying An Blevis and Eli Blevis, Photograph, 2020

Image 2.3 Only Tablet with Dark Hair (C), Shunying An Blevis and Eli Blevis, Photograph, 2020

We do this in the form of a poem and a collection of carefully posed and composed photographs—three here and 12 in the original. The poem is presented in English and in Putonghua (Mandarin Chinese). Here is the poem:

Night arrives, day recedes, I wake to meet my tablet's needs.
夕阳西下，夜幕降临，为了平板电脑上的需求，我辗转床榻
I do whatever Facebook wants, for fear of other people's taunts.
担心朋友反目，我听命于脸书
WeChat away th'entire day, "Moments" missed are not okay.
微信一整天，点遍朋友圈
Retweet whatever Twitter said, to show the world I'm less than dead.
为刷存在感，转推推特热点
I TikTok every night 'til two, tagging everything you do.
抖音过三更，分类分分钟
By my side throughout the day, at dinner, home, or when away.
随时随地，离身不能:吃饭，睡觉，居家，或旅行
Human here and AI there, into the dark lit night I stare.
一面是人情，一面是人工智能，我凝视着被照亮的虚空
Without a care, without a prayer, only tablet with dark hair.
其实没有人在乎，没有人祝福，只有平板电脑和黑发几缕

Image 2.4: Semi-invisible Mending, Shunying An Blevis and Eli Blevis, Photograph, 2021, Image 2.5: Invisible Mending, Shunying An Blevis and Eli Blevis, Photograph, 2021.

One strategy for sustainability is to make mending deliberately visible to highlight preservation, restoration, and reuse of existing materials as a matter of fashion. Mending is preferable to discarding things in favor of newly manufactured things.

Image 2.4 Semi-invisible Mending, Shunying An Blevis and Eli Blevis, Photograph, 2021

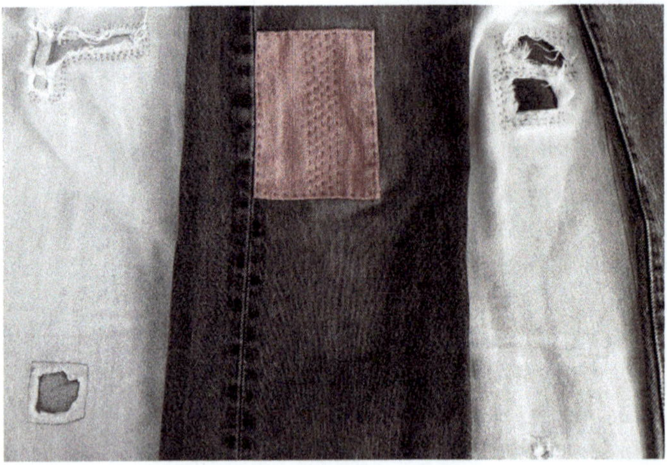

Image 2.5 Invisible Mending, Shunying An Blevis and Eli Blevis, Photograph, 2021

Image 2.6 Wisdom of Years, Shunying An Blevis and Eli Blevis, Photograph, 2017

Image 2.7 Foraged Foods, Shunying An Blevis and Eli Blevis, Photograph, 2017

Image 2.6: Wisdom of Years and Image 2.7: Foraged Foods.

In Wisdom of years (Blevis and Blevis, 2018), the authors provide an autoethnographic account of our parent's simple approach to a sustainable lifestyle. A lamp doubles as a stand on which to dry chilis from the garden. A variety of foraged foods are also documented, including black locust flowers that are edible, found mushrooms, and dandelions for tea. We also have a poem for this set. In this case, the original poem is in Putonghua. The English is a translation.

May our children be healthy and have jobs. May they have some savings in the bank. The characters on TV are easy to understand—they are either the good guys or the bad guys. Like all neighbors who are likeable. Eggs and chestnuts can be gifts. A smile is the friendliest language.

儿女要壮实，
有饭碗，
最好银行里有些存款。
电视里的人物很简单，
不是好人，
就是坏蛋。
街坊四邻都喜欢，
鸡蛋、栗子做贺礼，
微笑是最友好的语言。

Seasons change—there is no good reason for seasonal emotional changes. Will this year bring a good harvest or not? —that is my only concern. Real estate and equities are not long-term investments—rather, children. Species extinction and global warming are not a worry—rather, food on the table and heat in the winter.

不会伤春悲秋，
因为季节总在交替更换，
只讨论:
今年是否风调雨顺，
明年是否丰年。
不理股票、房产，
不担忧投资跌涨，
子女是他们的最大财产。
不担心种属灭绝，
也不担心全球变暖，
只在乎餐桌上有肉，
天冷了能取暖。

Life can be simple as it was several hundred years ago. Then, the air was good in our hometown. Then, there was no crazy house bubble, nor stock market boom and bust.

We say: Our parents' life is an arcadia. Happy and satisfied, our parents enjoy their lives. We hope this kind of living sustains eternally.

生活如此简单，
仿佛回到几百年前，
那时候没有雾霾，

Image 2.8 Petrol Station, Hong Kong (No Naked Lights), Shunying An Blevis and Eli Blevis, Photograph, 2012

没有大规模的环境污染，
没有房价飙升，
也没有股市崩盘。
身居闹市，
武陵桃源，
鸡犬相闻，
怡然自得享天年，
只希望能久远。

Image 2.8: Petrol Station, Hong Kong (No Naked Lights).

This petrol station is in the Hung Hom neighborhood of Hong Kong. The fossil fuels that it sells contribute to unsustainable conditions. The interaction design is in the signage, particularly. One part of one of the signs provides a mistranslation, “No Naked Lights.” If we persist in our use of fossil fuels, our entire world may figuratively go up in flames. Notwithstanding, replacing one infrastructure (e.g., petrol stations) with another (e.g., electric charging stations) is a complex task.

Image 2.9: Selfie Time

A sign near the Convention Center in Hong Kong proclaims, “Selfie Time.” The message echoes our preoccupation with technologies of self. The scaffolding below the sign advertises “China Evergrande,” a company that faced bankruptcy in 2024 because of its role in an unsustainable construction bubble in China. Another version of this photograph appears in Blevis, 2018.

Image 2.9 Selfie Time, Eli Blevis, Photograph, 2017

2.2 Summary

We have presented several images selected from several visually oriented publications connected to sustainable interaction design. The images here are chosen for quality and as examples of images in this space. Picturing sustainability as photographs in this way is a method for seeing and relating to sustainability in a way that gives meaning to conditions that would otherwise be only data. Some of the photographs we present here have been staged, and/or processed. Some are opportunistic. All are just a selection of moments that provide for reflection on sustainable and unsustainable ways of being. Over the years, we have published many such moments. We refer the reader/viewer to the references in this chapter for more photographs along these lines.

Acknowledgments Many thanks to the reviewers of the original pictorials and papers upon which this chapter is based.

References

Blevis, E. (2018). Photo Essay 1: Selfie Time. Photo Essay 2: Artifice and Nature. Photo Essay 3: Rooftop Garden. Photo Essay 4: Classroom Exercise. Photo Essay 5: Airstream. Photo Essay 6: Locked Gate. In M. Hazas & L. Nathan (Eds.), *Digital Technology and Sustainability: Engaging the Paradox* (p. 1,14,67,125,185,234). Routledge.

Blevis, E., & Blevis, S. A. (2018). Design Inspirations From the Wisdom of Years. In *Proceedings of the 2018 Designing Interactive Systems Conference (DIS '18)* (pp. 719–732). Association for Computing Machinery, New York, NY, USA. https://doi.org/10.1145/3196709.3196816

Blevis, S. A., & Blevis, E. (2022a). Only Tablet with Dark Hair (Pictorial). In G. Bruyns & H. Wei (Eds.), *With Design: Reinventing Design Modes. IASDR 2021*. Springer. https://doi.org/10.1007/978-981-19-4472-7_101

Blevis, S. A., & Blevis, E. (2022b). Visible Mending. In *Nordic Human-Computer Interaction Conference (NordiCHI '22)* (pp. 1–13). Association for Computing Machinery, New York, NY, USA, Article 87. https://doi.org/10.1145/3546155.3547291

Chapter 3
Seeing Robots Wearing Clothes

Zaiqiao Ye

Abstract The sight of robots wearing clothes can be a milestone in human–robot interaction. It serves as both a symbol of robots' integration into human society and human's embrace of robots as an indispensable part of their lives. This chapter offers a captivating visual journey through the world of companion robots adorned in diverse attires, drawing insights from an *X* (formerly *Twitter*) data research project and a robot workshop project. The *X* data research dives into the images generated by *X* users presenting their robot dog, AIBO, with clothes. The robot workshop project uses board games and drawings to help participants imagine their ideal companion robots. This chapter only showcases the selected visual outcomes of the two research.

3.1 Introduction

As robots become more visible in people's everyday environments, such as homes, schools, work, and entertainment spaces, it is not rare to see people dress up their robots or put accessories on them. Clothes provide a method for people to redesign or personalize their robots, thus potentially contributes to the sustainable use of robot product.

However, whether a robot should wear clothes remain a controversial topic (Friedman et al. 2021; Friedman et al. 2021; Hurtienne and Arnold 2020). The fact that people spontaneously clothe their companion robots inspired me to investigate what meaning and benefit robots' clothes provide for companion robot users. Clothing may be able to signal a robot's capabilities, help it fit into the social and cultural environment, or be a form of personalization for users (Friedman et al. 2021 Friedman et al. 2021). Investigating the use cases of robots' clothing and people's motivation to clothe their robots can be an inspiration for the future design of human-robot interaction for use in daily life.

The visual elements I used for this chapter is from two research:

Z. Ye (✉)
Shenzhen International School of Design, Harbin Institute of Technology, Shenzhen, China
e-mail: yezaiqiao@hit.edu.cn

S. An Blevis et al. (eds.), *Seeing Interaction Design*, Design Research Foundations,
https://doi.org/10.1007/978-3-032-14200-9_3

(a) A *X* (formally *Twitter*) data analysis study (Ye et al. 2023). The collected Tweets spans from 2019 to 2021. These tweets have images of AIBO decorated with clothes and accessories. The tweets are all open access and retrieved by R script using the academictwitteR package and then cleaned with JavaScript. This study falls into the category of non-human subjects research and do not need IRB review. For this chapter, I only use selected presentative images from the data collection process of the research.
(b) A series of workshops that utilized a board game to inspire participants to think about their life-long relationships with companion robots. The workshop centers around a board game designed by myself, called "Game of Life with companion robots" (see Image 3.10). The board game can help participants imagine about their future life with companion robots. During the workshop, participants used drawings and stickers to fill their storyboards. This study was approved by the university's institutional review board (IRB). For this chapter, I only use the drawings that contains participants' imagination of their personal robots' clothes.

In the following session, I provide the observation notes and design suggestions associated with the visual elements in this chapter. Moreover, I hope the visual elements themselves can inspire people to think about the field of robot clothes.

3.2 Observation Notes and Design Suggestions

(1) Consider clothes as a medium to provide emotional attachment and cultural meanings. The act of dressing up AIBO for special occasions like festivals and birthdays suggests that owners want to form emotional bonds with these robots (see Image 3.1). They may also want clothes to help companion robots to be better situated in some traditional events (see Images 3.4 and 3.6).
(2) The design of robots' clothes should consider their original form and function, and maybe provide additional value (see Image 3.2).
(3) Different sources of getting clothes for AIBO can have different outcomes in terms of sustainability. The DIY mode or repurposing other items can be more sustainable (see Image 3.3).
(4) A backpack looks fashionable on companion robots, meanwhile having the potential to provide some further functions (see Image 3.5).
(5) Clothes can signal companion robots' relationship with other robots (see Image 3.7).
(6) Clothes can signal companion robots' identity when they have group activities (see Image 3.8).
(7) Participants of the workshop do have imaginations about the clothes or accessories of their personal robots. Some participants even would like their companion robots to have a wardrobe (see Image 3.9).

(8) In a life-long relationship with companion robots, some people tend to use clothes to give an important role to the robots, such as a hero or mother (see Image 3.9).

3.3 Contribution

The chapter sheds light on the often-overlooked dimension of robot clothing. It contributes to both the field of human-robot interaction (HRI) and the intersection of Fashion and Sustainability. Through a visually engaging exploration, the visual materials inspire designers and researchers to think and rethink what role robot clothes play in our future lives, and how can we design future sustainable companion robots with the help of clothes.

Image 3.1 AIBOs are dressed up for Christmas, New Year's Day, and their "birthday"

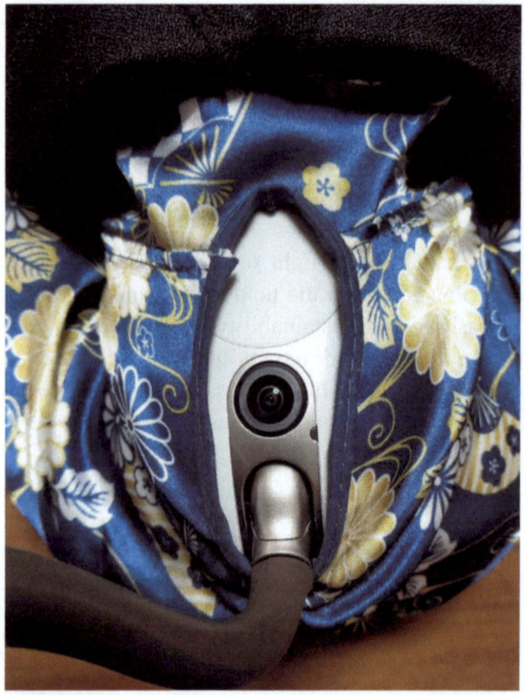

Image 3.2 The AIBO's suit is intentionally designed not to obscure the camera located on its back

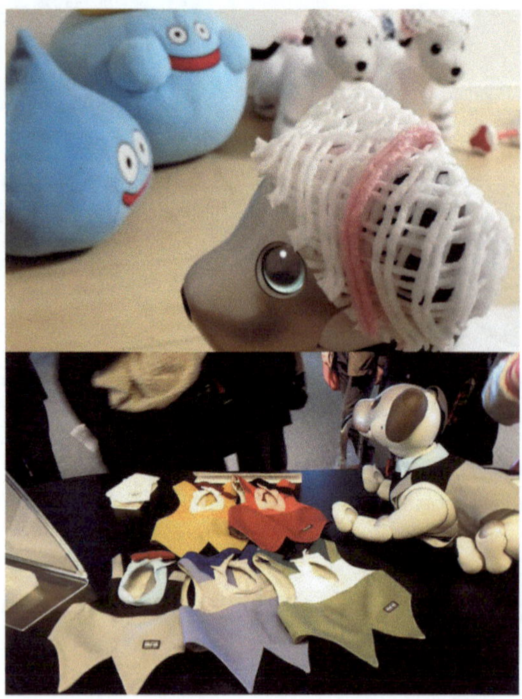

Image 3.3 Different ways that people can get clothes for AIBO. Upper: repurpose other item; bottom: buy them from small business

Image 3.4 AIBOs dressed up in Japanese traditional clothes

Image 3.5 Backpack as a fashionable item for AIBO

Image 3.6 AIBO in clothes doing a sacrificial ceremony

3.4 Summary

In this chapter, I first present a selective collection of pictures of AIBO in clothes provided by *X* (formally *Twitter*) users and approving public access. I select and group these pictures into various themes. Then I present a collection of participants' drawings regarding their imagination of robots with clothes in a workshop. Drawing from this visual collection, I have formulated some design recommendations for prospective human-robot interaction designers to consider and explore. Meanwhile, I hope the visual representations can enhance our comprehension of companion robots' clothing choices and their presence in real-life settings.

Image 3.7 AIBOs in pair wearing clothes

Image 3.8 AIBOs in group wearing clothes

Image 3.9 Participants' Imagination of Their Personal Robots' Clothes

Image 3.10 The board game "Game of Life with your companion robot" and the workshop setting

Acknowledgements I extend my gratitude to my previous colleague Wei-chu Chen for writing the R script to retrieve the *X* (formally *Twitter*) data. I also want to give special thanks to the X users and the participants of the workshop. Your contributions were essential to this research.

References

Friedman, N., Love, K., Bremers, A., Parry, A. J., Ray, L. C., Bolor, A., Jen, L., & Wendy, J. 2021. Designing Functional Clothing for Human-robot Interaction. In *Companion of the 2021 ACM/IEEE International Conference on Human-Robot Interaction (HRI '21 Companion). Association for Computing Machinery, New York, NY, USA*, 703–705. https://doi-org.proxyiub.uits.iu.edu/10.1145/3434074.3444870

Friedman, N., Love, K., Ray, L. C., Jenny, E. S., Guy, H., & Wendy, J. 2021. What Robots Need From Clothing. In *Proceedings of the 2021 ACM Designing Interactive Systems Conference (DIS '21). Association for Computing Machinery, New York, NY, USA*, 1345–1355. https://doi-org.proxyiub.uits.iu.edu/10.1145/3461778.3462045

Hurtienne, J., & Arnold, D. 2020. The Naked Truth? In *Companion of the 2020 ACM/IEEE International Conference on Human-Robot Interaction (HRI '20). Association for Computing Machinery, New York, NY, USA*, 269–271. https://doi-org.proxyiub.uits.iu.edu/10.1145/3371382.3378362

Ye, Z., Chen, W. C., & Sabanovic, S. 2023. Dressing up AIBO: An Exploration of User-generated Content of Robot Clothing on Twitter. In *Proceedings of the 11th International Conference on Human-Agent Interaction (HAI '23). Association for Computing Machinery, New York, NY, USA*, 81–89. https://doi-org.proxyiub.uits.iu.edu/10.1145/3623809.3623852

Chapter 4
Conversational Composites: Unexpected Layers

Mafalda Gamboa, Miriam Sturdee, and Sara Ljungblad

Abstract The conversational nature of sketches is a widespread topic of research. Understanding drawing as a cognitive activity is commonly accepted, and many of the most extensively used methods within Human–Computer Interaction recruit sketching as a technique for ideation, explanation, documentation, and conversation. To further develop the use of this illustration process as a tool of knowledge production, we suggested a novel sketching method. We presented *Conversational Composites*: a flexible method grounded in the material and tangible qualities of sketching in different forms and media, creating physical and digital layers of conversation between participants. We introduced and reflected on the proposed method through an applied case of a conversation between a PhD student and her supervisor. In this pictorial, the conversation is extended, and we discuss an unexpected case of when our method was introduced at a conference—specifically how biomaterials (SCOBY) started sketching and living with us. We also reflect on how layering of research and practice between projects can result in new learning.

4.1 Introduction

The nature of drawing—in this case sketching and illustration—is dependent on the tangible characteristics of the material used. The media defines the possible vocabulary and determines the spectrum of afforded drawing conversations. For example, drawing on a white board with a pen is entirely different from drawing on paper with brush and ink. We have previously introduced *Conversational Composites*:

M. Gamboa · S. Ljungblad
Chalmers University of Technology, Goteborg, Sweden
e-mail: mafalda.gamboa@chalmers.se

S. Ljungblad
e-mail: sara.ljungblad@chalmers.se

M. Sturdee (✉)
University of St Andrews, St Andrews, UK
e-mail: ms535@st-andrews.ac.uk

S. An Blevis et al. (eds.), *Seeing Interaction Design*, Design Research Foundations,
https://doi.org/10.1007/978-3-032-14200-9_4

a method inspired by the exquisite corpse technique, but with a twist: participants build upon earlier fragments transparently using different drawing media layers, fostering illustrated dialogues (Gamboa et al., 2023). This approach complemented verbal communication in group discussions, allowing for subjective expression and engagement with design practice.

Sketching, whether in the form of comedic pieces, programming code, technical diagrams, or artistic images like watercolor paintings, serves as a versatile tool in technological inquiry, enabling low-fidelity design and multi-modal expression. Rooted in visual communication and inviting responses, sketching facilitates dialogue among researchers, designers, and artists, fostering unexpected insights and inspiring imagery through iterative and collaborative processes.

The method we proposed involves creating a sequence of layers through contrasting artistic characteristics to build on each other's ideas, leading to speculative designs. It draws inspiration from Yurman's Drawing Conversations (Yurman & Reddy, 2022), which uses watercolor to facilitate collaborative visual dialogues among participants. This method differs from existing approaches such as the *exquisite corpse* technique by using original sketch-based and visual forms, applying layering for reinterpretation and ambiguity. Unlike other methods, *Conversational Composites* focuses on building narratives with less speculation and more emphasis on reinterpretation. While the techniques used are not new individually, their blending creates a new design space applicable to various topics, offering a novel perspective on design and research issues through layered narrative prototyping. Image 1 shows an overview of the original process.

4.1.1 Revisiting the Conversational Composites—A New Layer

The original conversation took place between Sara and Mafalda, iterating upon 'micro-stories' they had collected from drone pilots. When I (Miriam) was invited to join the project, the conversations were finished, the layers dry and complete. My role was to interpret the technique in terms of contemporary literature, and consider the method and its role in research, its future possibilities. This brought with it a sadness of not being part of the original process, but I was none-the-less fascinated to see how each artist-researcher had brought their own style and thought process to the topic.

To continue the conversation, I created digital prints of the final layer of each of six composites, moving from organic material (tracing paper) to plastic transparencies and permanent marker, in contrast to the original fluid approach using watercolour paints, which then transitioned via tracing paper, to Procreate on iPad. This process felt like viewing the images through a lens, marking-up and analysing rather than letting my imagination wander. However, I realized that there were still new insights, both in terms of process and in reinterpretation, the black marker allowing me to pinpoint particular areas of interest, and reinterpret or overlay others.

Images 4.2–4.6 show the steps taken to produce the original conversational composites, each with three examples of the emerging narratives as the layering took place, with one additional 'post-production' reflective layer. Each original 'micro-story' is included in the capton, with the new reflections in the image commentary.

4.1.2 Layers in Research & Presentation

To present the original work, we (Miriam and Mafalda) travelled to Poland to attend a conference on Tangible, Embedded, and Embodied Interfaces (ACM TEI). It was a small conference where we still got to meet several interesting researchers. We encountered Fiona Bell and Eldy S. Lazaro Vasquez and their collaborators. During the first day of the conference, Mafalda participated in their workshop on exploring SCOBY (Symbiotic Culture of Bacteria and Yeast) as a material for HCI (Bell et al., 2023). Their work was also presented at the conference, in the form of an interactive corset. They had kindly brought with them a few sheets for us to play with during the workshop. SCOBY is a sustainable biofilm grown in kombucha tea, that when dried feels and acts similarly to leather. It has a distinct smell, and their SCOBY sheets had also been mixed with lavender oil, so their scent was quite pungent. Two things happened: we began to use this SCOBY as a drawing material and Mafalda used it as a screen cover for her phone for the whole duration of the conference.

In the words of Tim Ingold in Correspondences: "*Ash is the shadow of a woodfire, a collapsed mound of wax the shadow of a candle, pitch the shadow of pine. Grease is the shadow of fingers and sweaty palms on the tools we use, on the handles with which we open doors, and the turned pages of well-thumbed books. Ash, wax, pitch, and grease: these shadows have substance. They are not projections of rigid form but the residues of material flows, admixtures and mutations.*" (Ingold, 2020). SCOBY introduced these unexpected dimensions to our Conversational Composites.

4.1.3 SCOBY as a Sketching Material

Inspired by this biomaterial and its capacity to be sketched on, adding a layer of patina to our work, we decided to hold a small exercise during the presentation of our pictorial. We asked the audience to draw on paper, two layers of translucent paper, and finally one layer of SCOBY. The audience thought the exercise was fun, but more interestingly, there seemed to be a deep engagement with understanding the qualities of the last material by touching it, smelling it, and observing how its texture contributed to the composite of drawings. Many commented on how the threads from the bacteria seemed to be lines in the drawings.

4.1.4 Layering SCOBY on a Phone

The layering of the SCOBY worked as a material beyond drawing media. We found out that not only we could draw on the SCOBY, but also that it could be used as a layer on top of a touchscreen. I (Mafalda) cut a small piece of the SCOBY and shaped it to work as a screen protector for my phone and promised the workshop organisers to keep it there for the duration of the conference. This turned out to be a more engaging experiment than initially expected. A few things happened. First, people around me reacted at the sight of a phone with a textured yellow screen. But I could have not expected my own reaction to the second skin: when I woke up the next morning and reached out for my phone to turn off the alarm, the stench was so poignant I promptly put the phone away from my face. This was quite refreshing, as I never found my morning use of the phone anything worth preserving. I kept the layer until returning home, where it ended up scrunched up on the floor, peeled off by one of my curious children. In this exchange, I re-evaluated how a layer upon a digital screen could be yet another facet of the composite. By drawing on my iPad Pro and using a SCOBY layer, the drawings lost much of their fidelity.

4.2 Contributions

The method's application sparked discussions on novel research avenues. Surprisingly, the silent deliberations proved invigorating, offering visual insights into the significance attributed to each depicted element. For instance, Mafalda, the PhD Student, discerned a pronounced focus on non-human aspects within human-drone interaction, prompting a reconsideration of research questions and stakeholder inclusivity, such as pondering the involvement of trees and birds or marine life when drones malfunction in water. Sara, an adept watercolorist, found the approach delightful, intuitive, and playful, facilitating collaborative reflection on user data in a unique manner distinct from conventional qualitative analysis methods like thematic analysis or coding.

The method's adaptability and reliance on physical media suggest limitations for remote application, yet its versatility in accommodating various prompts, tools, and participants indicates promise. Future endeavors may entail compiling a repository of procedural steps with examples and evaluating the method's efficacy across diverse participant cohorts. The tangible nature of the composites, each layer inviting distinct artistic expression, renders them invaluable artifacts fostering interdisciplinary engagement and dialogue, notwithstanding challenges like interpretational ambiguity and annotation facilitation. Although not every composite may yield generative outcomes, leveraging them as imagery for interdisciplinary discourse underscores their utility.

The additional layering emphasizes the longevity of the materials, in inspiring or provoking further interpretation and enquiry by revisiting visual data. Textual

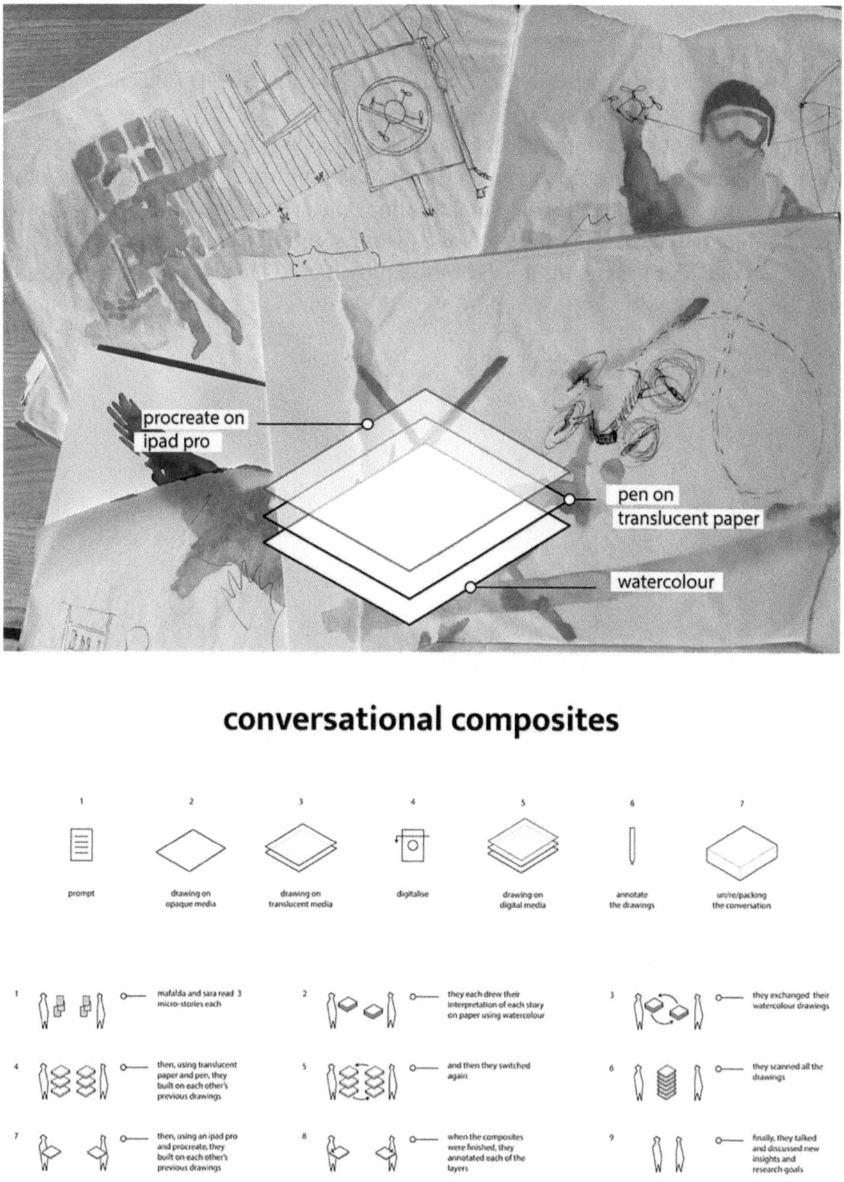

Image 4.1 The conversational composites technique, illustrated

data is often reinterpreted, quoted, and woven into further narratives, but we rarely examine visual data sets under a new lens. Each new layer both continued the original dialogue, but also was found to garner new insights.

At the time of presentation, the method's evolution also beckoned further exploration and systematic reflection. The conference afforded an opportunity for collective growth and refinement, which we jumped upon by superimposing our concept with an available biomaterial. This was a material we had never considered as a potential layer, and it brought visual dimensions and an influence of bacteria as an agent in the drawing composite. Not only that, but the SCOBY became another interactive layer superimposed to the digital screen.

Image 4.1 The Conversational Composites Technique, Illustrated

The original technique, a composite image showing the layers shuffled together with diagrammatical information which outlines the steps taken both for the hands-on technique, but also of the research as a whole—prior to the addition of the new layers—an outsider looking in. Each of the following images is named after a quote from the micro-story it was inspired by, with the original pictorial paper reference in brackets.

Image 4.2 'It Flew Away' (A)

Image 4.2 "It flew away. I used the app litchi and had created a route it would follow, but it never came back. It turned out that I did not think that the stated height is in relation to the starting point and when the ground rose, it came closer and closer to the ground and finally stopped in front of a large spruce and did not know where it would go. There it hovered until the batteries ran out. A resident in the area found it after 1 1/2 years and when I checked the film on the memory card it was clear what had happened"

The original layers focused on the prominence of the forest and a single spruce, looking sadly upon a cemetery of missing drones. In this new layer, I focus not on the spruce trees as friendly bystanders, but active participants in the destruction of drone technology. Claw-like branches reach out and snag the drones from the sky. I added tree stumps to show the mismatch between human destruction of the forest and the loss of individual pieces of technology. There is a tension when introducing technology into the natural world, that we have already taken so much from. Whereas the original paper considers nature as a stakeholder, here it is considered a victim, trying to preserve the natural order.

Image 4.3 'Unpleasant Experiences' (B)

The original layered image deliberately obscured the first two layers with the addition of the digital layer, to emphasise the obfuscation of potential problems for the drone pilot. Much of the meaning of the original layers was misinterpreted, but this only emphasised the concept of 'hidden' objects. I felt that drones may be

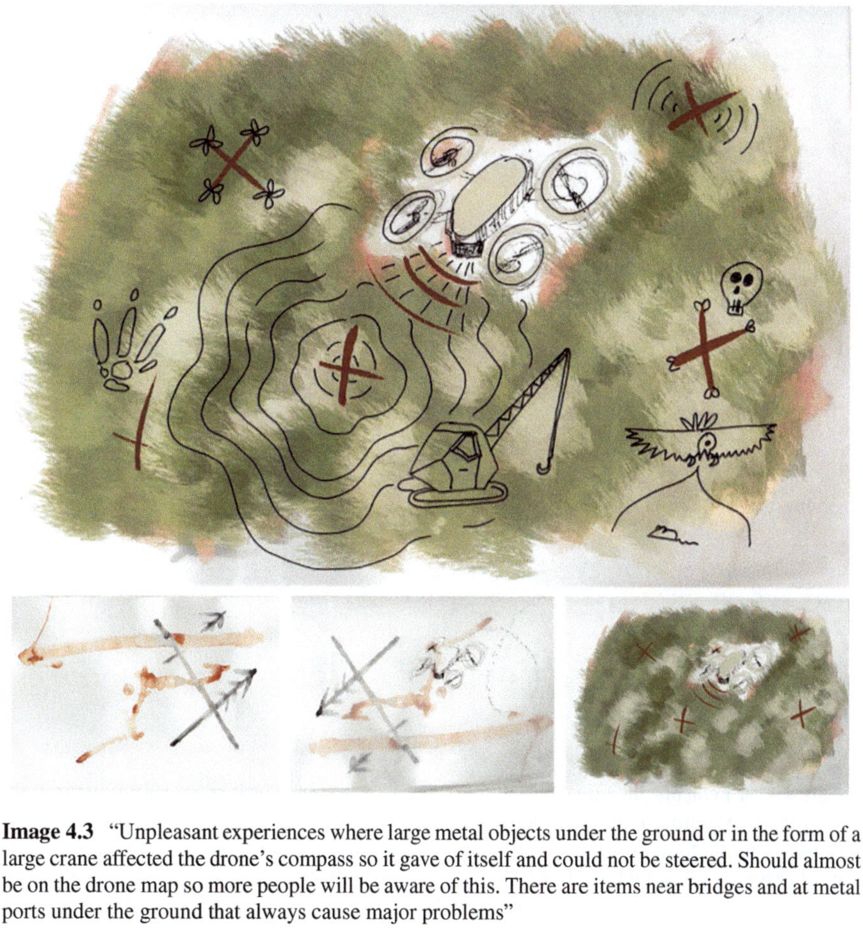

Image 4.3 "Unpleasant experiences where large metal objects under the ground or in the form of a large crane affected the drone's compass so it gave of itself and could not be steered. Should almost be on the drone map so more people will be aware of this. There are items near bridges and at metal ports under the ground that always cause major problems"

influenced by magnetic fields from buried metal objects, but they also are used in land surveys, to discover what lies beneath. The story also emphasised an unburied crane, so I added this in, showing the land being worked. The bird in the bottom right corner shows that the aerial view is vital for discovering food sources, and bottom right I added a dinosaur footprint to hint that there is more to discover in the land once you look at it from a different viewpoint. I annotated the red crosses added by Mafalda in the previous layer to show obstacles, showing and danger and even a buried drone.

Image 4.4 'I'm on My Fifth Drone' (C)

I was struck by the nuances of the imagery in the previous layers, there were so many threads—a concept that Mafalda touched upon in the second layer in reference to the kite-surfing kit. I wanted to connect all of the different concepts in the story

Image 4.4 "I'm on my 5th drone. I have crashed three drones and that is of course due to the way I use the drone. My drone is a camera dolly and the best movie clip is when you drive backwards and sideways. Unfortunately, you do not see in the direction of travel either. I usually film my grandson, who engages in kite surfing and kite foiling, 2-3 m above the water is usually the best and safest, but sometimes you end up below 1 m and then there is a crisis"

that looked like they used this metaphor. The threads of seaweed, the lines of the kite, the breathing tube of the diver. I thought of the other lost drones and how all of the incidents were connected, so I added a ball of wool, additional drones, a thread between them, then a piece of woven cloth to show that these are all part of the same woven story.

Image 4.5 'Waiting to Put It's Claws On' (D)

Nature was the antagonist in the original series. The drone pilot was very protective of their new technology and was deterred by the circling hawk. This is at odds with Image 4.3, where originally the trees were seen as benevolent or sad protectors and I suggested taking action against technology. Again I annotated the natural, but this time the technology becomes merged with the animal. The bird is now an aircraft in itself, with fixed wings and cargo bay doors to release missiles. Rather than a near miss, the drone is lying broken, bleeding oil. I also annotated the person in the

Image 4.5 "One fine spring day I was going out with my newly bought drone. Just as I am about to take off and fly away with the drone, I see a large eagle just above me circulating around me. Almost like it was waiting to put its claws on. Do you think it was a ride after that shock? Nope!"

bottom corner to show that they are protecting their 'bubble'—the drone becomes the vulnerable party.

Image 4.6 'Everything is On Film' (E)

This image was so beautiful I was reluctant to add to it. The story itself is a repeating tale of a drone colliding with a tree, but the context of the forest and bridge and the changing seasons was the most compelling part. The way Mafalda and Sara had composed the layers looked like a page from an illustrated children's book. They emphasised the natural landscape, and here, despite the drone getting 'bounced around in the birch', there is less of a clash with nature, as the pilot is trying to capture the beauty of the natural world, rather than being frustrated by its existence.

Image 4.6 "Meant to lift from a table on a small bridge. Had run the same sequence last autumn but now it was winter and wanted the same bit in winter but did not want footprints in the snow so flew into the place. When I took off I hit a branch and the drone bounced around in a birch… the propellers broke partly but managed to steer it down with difficulty and crashed on the small bridge and not in the water. Everything is on film"

I added a small silhouette of the pilot in the background, and a snowman to represent the playful, storybook style, adding a small bird in the tree, apparently unperturbed by the presence of technology. On the far right I sketched a rip to show that there was a purposeful divide between seasons, but the rip is organic and not informed by human interference.

Image 4.7 'An Angry Man Came' (F)

This conversational composite felt quite simple, the man is angry and does not want the drone flying over his property. Mafalda added some signage, which clearly shows the intent of the person with their arms outspread—no drones! I found it strange that neither Mafalda or Sara had focused on the other part of the story—that there had previously been a family who were interested. Here, I merge the family and the farmer, they are sad he is angry. I also fast forward and imagine the drone has been shot out of the sky, and the sign has become a memorial for the fallen drone. The outstretched arms of the farmer also appeared to me that he was trying to flap his arms and take off, so I added movement lines. I changed the nature of the story to

Image 4.7 "A family became interested when I flew my drone before. At about the same place a few months later, an angry man came and said that he would shoot down the drone if I flew over his house"

Image 4.8 Creating conversational composite materials using SCOBY sheets; inset—conference attendees reflect upon the collected imagery

one of wistful envy, but a desire to be the only one who is allowed to see the birds-eye view of the property.

Image 4.8 Creating Conversational Composite Materials

Mafalda was given SCOBY sheets to take away from the workshop, and we decided to create mini-conversational composite kits to hand out to the attendees of

Image 4.9 Three iterations of SCOBY: discarded pieces during the workshop, reclaimed SCOBY as part of the conversation, finished conversational bio-composites

our planned talk. Originally, we were ambitious, purchasing multiple sheets of paper and tracing paper, hoping to create dense layering of sketches, but when the layers become too dense, it is impossible to see what has come before. The SCOBY, as a natural material, also is non-uniform, offering its own interpretation of what lies beneath.

Image 4.9 Three Iterations of Scoby

The workshop used multiple sheets of SCOBY, many of which were destroyed in the playful making that took place. Later, the surviving pieces were used in a different way, to build conversational composites about the presentation we gave—composites about composites. After the presentation, the layers were shared and discussed informally, showing both inspiration from the talk itself, but also from the imagination of the attendees, and their reactions to the biomaterial and layering.

Image 4.10 SCOBY as Screen / SCOBY as Refuse

Having already demonstrated it was possible to use SCOBY as a phone screen, Mafalda carefully took home an additional sheet to explore this concept further. Much as with the original conversational composites, she finished with digital sketching on the iPad using Procreate, but tracing *through* rather than *on*. The uneven surface and

Image 4.10 Top—SCOBY as tracing paper over digital tablet; Bottom—the discarded SCOBY crumpled from use becomes refuse

colour appears as a combination of tracing paper and screen—at odds with the drone stories, here nature and technology work together—and finally, unlike the missing drones in lakes or forests, the discarded SCOBY will biodegrade.

4.3 Summary

By bringing together layers, upon which we can leave traces, conversations happen. The threads that run between the images focus on a complex interplay between people, technology, and nature. There is both the potential to continue layering to discover new insights, but also to saturate understanding, by obscuring what has come before. Interrogating and exploring imagery, materiality and interaction can also be seen as a method of participation between groups, and storytelling to inform research and practice. Conversational Composites offer a methodology for both subjective and multiple insights.

Acknowledgements With thanks to Fiona Bell, Eldy S. Lazaro Vasquez and collaborators for their engagement with the researchers, and for providing the SCOBY to use as inspiration for sketching. Additional thanks to Fiona Bell and the workshop photographer for supplying additional imagery for use within this pictorial.

References

Bell, F., Chow, D., Lazaro Vasquez, E. S., Devendorf, L., & Alistar, M. (2023, February). Designing Interactions with Kombucha SCOBY. In *Proceedings of the Seventeenth International Conference on Tangible, Embedded, and Embodied Interaction* (pp. 1–5).

Gamboa, M., Ljungblad, S., & Sturdee, M. (2023, February). Conversational Composites: A Method for Illustration Layering. In *Proceedings of the Seventeenth International Conference on Tangible, Embedded, and Embodied Interaction* (pp. 1–13).

Ingold, T. (2020). *Correspondences.* John Wiley & Sons.

Yurman, P., & Reddy, A.V. (2022, June). Drawing Conversations Mediated by AI. In *Proceedings of the 14th Conference on Creativity and Cognition* (pp. 56–70).

Part II
Seeing and Noticing

Chapter 5
All the Tea in China, Seeing What Is and What Can Be, Visual Thinking Gallery, Being Photo Visual

Shunying An Blevis and Eli Blevis

Abstract The idea of *seeing and noticing* is not unique to interaction design. Notwithstanding, inasmuch as *seeing and noticing* are foundational skills for any engagement with Design, *seeing and noticing* are also foundational skills for interaction design. In this chapter, we have selected several images from the papers and pictorials listed above. The selection is based on sampling this work but does not replace the original work. Moreover, in including images and descriptions here, we are illustrating selected aspects of seeing and noticing that are visually compelling and thought provoking.

5.1 Contributions

We give examples of seeing and noticing in several contexts, from (a) the effects of a drone appearing above a crowded performance in Macau, to (b) street life in Hong Kong that relate to distribution and recycling and on closer inspection to respect for work and implications for sustainability, to (c) to respect for work in relation to wealth inequity, to (d) respect for work in relation to "migrant" domestic workers, to (e) interaction design and tea in relation also to respect for work.

Image 5.1: Drone, Eli Blevis, Annotated Photograph, 2015, Image 5.2: Drone, Eli Blevis, Annotated Photograph Details, 2015.

Photographs in this chapter are selected from *All the Tea in China (with Bonnie Nardi)* (Blevis et al., 2019), *Seeing What is and What Can be* (Blevis, 2018), *Visual Thinking Gallery* (Blevis, 2016a), *Being Photo Visual in HCI and Design* (Blevis, 2016b), and *Drones* (Blevis, 2015)

S. A. Blevis · E. Blevis (✉)
Luddy School of Informatics, Computing, and Engineering, Indiana University at Bloomington, Bloomington, USA
e-mail: eblevis@iu.edu

S. A. Blevis
e-mail: san@iu.edu

S. An Blevis et al. (eds.), *Seeing Interaction Design*, Design Research Foundations,
https://doi.org/10.1007/978-3-032-14200-9_5

Image 5.1 Drone, Eli Blevis, Annotated Photograph, 2015

In an old section of Macau, a large crowd watches a local dance performance. Many people are watching and photographing or making videos of the performance. Above the crowd, a drone is flying. The attention of some of the audience for the dance performance is perturbed by the presence of the drone. These few audience participants are looking up at the drone, rather than the performance. The photographs

Image 5.2 Drone, Eli Blevis, Annotated Photograph Details, 2015

are annotated similarly to a notion of annotation introduced in interaction design and HCI by Gaver and Bowers (2012).

The drone operator is not visible to the crowd. He is perched on a hill above and out of sight of the performance venue. The many photographers and videographers line the front row of the audience. The performers can see their faces from behind their cameras, but not the drone operator. The performance was a variety show with different local acts. The act pictured is of a young couple dancing—a man in a top hat and sleeveless shirt, a woman in high heel shoes and a sleeveless dress with ample twirl.

These photographs appeared in Blevis (2016b), in a full-length article and in Blevis (2015) as a one-page photograph on the inside cover of ACM *interactions*. In 2015, the full impact of flying drones over crowds of people was not fully appreciated, it seems. The 2015 photograph at the time we write this has 95 citations per semant icscholar.org, mostly in articles concerned with military and security implications of drones.

Image 5.3: Distribution, Eli Blevis, Photograph, 2007 (1/3), Image 5.4: Distribution, Eli Blevis, Photograph, 2007 (2/3), Image 5.5: Distribution, Eli Blevis, Photograph, 2007 (3/3), Image 5.7: Distribution, Annotated, Eli Blevis, Annotated Photograph, 2007.

In an intersection of a historic district Lan Kwai Fong (Sheung Wan, Hong Kong) with narrow streets & alleys, an elderly delivery man (A) uses his body to keep his over-loaded delivery cart from slipping. In the center of the down-hill street, there are big STOP signs (G) in white in both English and Chinese to warn approaching traffic to use brakes from a distance. The elderly delivery man seems very experienced judging from how he lows his center of gravity. However, manually maneuvering an over-loaded delivery cart on a steep hill proves to be physically challenging. His struggle catches the attention of a nearby pedestrian (B).

As e-commerce booms, more and more young people are attracted to the delivery industry. In the west, recently some delivery drivers are getting competitive compensations and improved working conditions. In Asian countries, such as China, the high-profit and fast-growing e-commerce still relies heavily on delivery service that

Image 5.3 Distribution, Eli Blevis, Photograph, 2007 (1/3)

Image 5.4 Distribution, Eli Blevis, Photograph, 2007 (2/3)

Image 5.5 Distribution, Eli Blevis, Photograph, 2007 (3/3)

is done by bicycle, cart, and motorbike. These delivery workers often work long hours under challenging working conditions, not to mention the demanding customers (c.f., Feng, 2020)

Image 5.6: Recycling, Eli Blevis, Photograph, 2007, Image 5.8: Recycling, Annotated, Eli Blevis, Annotated Photograph, 2007.

Image 5.6 Recycling, Eli Blevis, Photograph, 2007

Image 5.7 Distribution, Annotated, Eli Blevis, Annotated Photograph, 2007

In a busy street of Mong Kok (Hong Kong) knowing for shopping, an elderly man (A) is collecting most any form of cardboard and other materials for recycling on his car-length cart. The cart is so overloaded that even as the man tries to add the two poles in his hand, some of the treasure (B) is falling off the front. Nearly all the cart's contents appear to be packaging. One of the packages was from a Japanese manufactured vacuum cleaner (D). Many of the packages are from Brazil—*Penasul*

Image 5.8 Recycling, Annotated, Eli Blevis, Annotated Photograph, 2007

Image 5.9 The Bentley on the Terrace, Daybreak, Eli Blevis, Photograph, 2016

(E) is a Brazilian food company that exports chicken feet to Hong Kong. Brazil is not very close to China. There are environmental costs to using non-local products.

The man driving the car (C) immediately in the background behind the man with the cart is honking his horn—cars in Hong Kong are right-hand drive. The cart is in the middle of the street. It is blocking the car. The other side of distribution, after we have consumed things, is the salvage, disposal, or recycling of the parts of products

Image 5.10 The Bentley on the Terrace, Midday, Eli Blevis, Photograph, 2016

Image 5.11 The Bentley on the Terrace, Evening, Eli Blevis, Photograph, 2016

that we did not use. The value of this elderly man's work is in the collection of these packaging materials for recycling. The value of these materials compared to their size must be quite small. So, the man must collect as much as he can for each trip. This need creates his endlessly repeating Sisyphean task. This kind of labor-intensive and low-paid work is often done by elderly people living in poverty in Hong Kong.

Image 5.12 CBD, Migrant Domestic Workers Day Off, Eli Blevis, Photograph, 2014

Image 5.13 Tea, Shunying An Blevis and Eli Blevis, Photograph, 2019

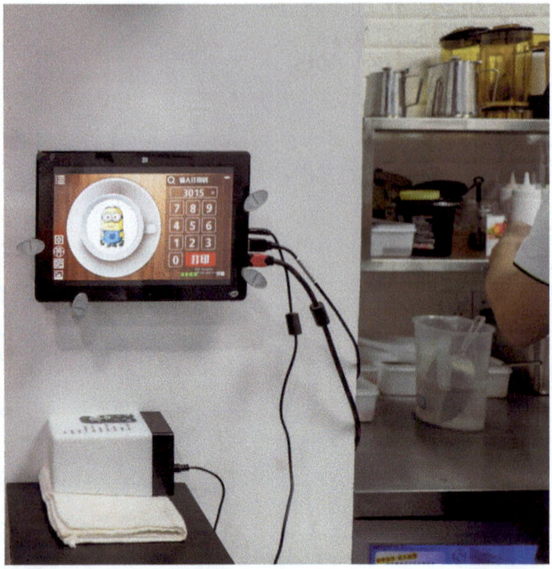

Image 5.14 3D Printed Tea—Order Tablet, Shunying An Blevis and Eli Blevis, Photograph, 2019

Image 5.15 3D Printed Tea—Served, Shunying An Blevis and Eli Blevis, Photograph, 2019

The impatience of the man in the car seems misplaced, given how everyone who consumes packaged products relies on this man's work.

Keegan (2018) takes up the question of should this elderly man be paid more, and be paid with more respect?

Image 5.16 3D Printed Tea, Recycling, Shunying An Blevis and Eli Blevis, Photograph, 2019

Image 5.17 3D Printed Tea, More Recycling, Shunying An Blevis and Eli Blevis, Photograph, 2019

Image 5.9: The Bentley on the Terrace, Daybreak, Eli Blevis, Photograph, 2016, Image 5.10: The Bentley on the Terrace, Midday, Eli Blevis, Photograph, 2016, Image 5.11: The Bentley on the Terrace, Evening, Eli Blevis, Photograph, 2016.

The three photographs were taken from a high rooftop on the terrace of Hong Kong Coliseum. They appeared in Blevis (2016) as a visual essay with a group of other photographs, depicting the activities on the terrace in 1 day. These three photographs are the beginning, the middle, and the end of the sequence. In order, the sequence starts with caretakers strolling empty carts to the terrace. We can see that these carts are empty in the relaxed posture of the caretakers. In the midday, the Bentley appears. Finally, the caretakers in the first image appear at the end of their day. Their posture indicates that their carts are full and heavy with things to be removed from the terrace. The photographs that depict the full day buzzing activities of caretaking the terrace and a sword dancing class are not included in this chapter.

The caretakers in these photographs are maintaining the landscape and removing waste manually outdoors in the heat on the terrace. The Bentley that suddenly appears at noon seems out of place, especially in contrast to the lifestyles of the caretakers.

Image 5.12: CBD, Migrant Domestic Workers Day Off, Eli Blevis, Photograph, 2015.

Migrant domestic workers occupy an outdoor stairway in the Central Business District of Hong Kong on a Sunday, their day off, in May 2014. We can surmise that this social gathering occurs here because there are not enough more suitable locations available for this population.

There is an escalator on the side of the staircase. A young woman on the escalator stares back at the camera, as if it is the photographer's interest in this scene that is unusual. The photograph is in black and white, because it was taken with a digital camera equipped only with a monochrome sensor.

Image 5.13: Tea, Shunying An Blevis and Eli Blevis, Photograph, 2019.

Three women show off fine China teacups newly gifted as heirloom objects from their elderly relative on a roof top terrace in Toronto. These teacups were once finely crafted and luxury possessions. They may have served many social and relaxing gatherings. They will continue to support such fine tea-drinking occasions in their new owners' life.

Image 5.14: 3D Printed Tea--Order Tablet, Shunying An Blevis and Eli Blevis, Photograph, 2019, Image 5.15: 3D Printed Tea—Served, Shunying An Blevis and Eli Blevis, Photograph, 2019.

In a world where innovation is considered imperative, "stuck in the past" is a fear of many in traditional industry. What do contemporary digital technologies offer to the centuries-old traditions of engagement with tea? The Chá Lî Chá Lî tea shop offers customers a chance to select or upload images which may be printed on yogurt and tea-based drinks. Although called "3D," this is really a 2D print on top of the beverage. The images show the ordering tablet, the "3D" printer, the drink as delivered in a disposable container with two plastic lids and a plastic straw wrapped in plastic, and the effect of drinking the tea on the image of the minion in this instance.

By combining food safe printing, mobile connectivity, shared tablet interface, and customer images or stock cartoons, the tea drinking experience can be endowed with novelty and fun. Novelty and fun enabled by novel combinations of new interactive technologies can also be frivolous.

Image 5.16: 3D Printed Tea, Recycling, Shunying An Blevis and Eli Blevis, Photograph, 2019, Image 5.17: 3D Printed Tea, More Recycling, Shunying An Blevis and Eli Blevis, Photograph, 2019.

One hundred eighty degrees opposite the Chá Lî Chá Lî tea shop, two women sort through trash to collect recyclable materials in the first of these two images. In the background upper right, we can see some young lovers on a bench. A young man walks by preoccupied by his mobile. A few steps away from the women in the first image, a man engages in the same occupation in the second of these images. His blue mini truck is parked behind him. Even when the effects of consumption of disposables is clearly in view and close at hand, we consumers may still be blind.

The photographs here are street Art photography. The wide angle of view is key to showing context. As an interaction design inspiration, we propose: Consider all effects of every design. Look in every direction to see all stakeholders.

5.2 Summary

In this chapter and as we state in the section titled *Contributions*, we give examples of seeing and noticing in several contexts, from (a) the effects of a drone appearing above a crowded performance in Macau, to (b) street life in Hong Kong that relate to distribution and recycling and on closer inspection to respect for work and implications for sustainability, to (c) to respect for work in relation to wealth inequity, to (d) respect for work in relation to "migrant" domestic workers, to (e) interaction design and tea in relation also to respect for work. Herein, *Seeing and Noticing* are not just acts of photographic street Art, but also techno-political observations. This is not the gaze of an ethnographer writing the culture. Rather, many of these images serve as examples of everyday inequalities that may be observed in the day-to-day use of drones, carts, stairs, printers, and packaging.

Acknowledgments Many thanks to the reviewers of the original pictorials and papers upon which this chapter is based.

References

Blevis, E. (2015). Drones. *Interactions, 22*(5), 80.

Blevis, E. (2016a). The visual thinking gallery: A five-year retrospective. In *Proceedings of the 2016 ACM conference on designing interactive systems (DIS '16)* (pp. 1096–1110). ACM.

Blevis, E. (2016b). Being photo-visual in HCI and design. In *Proceedings of the 2016 ACM conference on designing interactive systems (DIS '16)* (pp. 983–995). ACM.

Blevis, E. (2018). Seeing what is and what can be: On sustainability, respect for work, and Design for Respect. In *Proceedings of the SIGCHI conference on human factors in computing systems (CHI '18)*. ACM.

Blevis, S. A., Blevis, E., & Nardi, B. (2019). All the tea in China: Interaction design inspirations. In *Proceedings of the 2019 on creativity and cognition (C&C '19)* (pp. 333–345). ACM.

Feng, E. (2020, December). *For China's overburdened delivery workers, the customer—and app—is always right. All things considered*. NPR. Retrieved January 1, 2024

from https://www.npr.org/2020/12/01/938876826/for-chinas-overburdened-delivery-workers-the-customer-and-app-is-always-right
Gaver, B., & Bowers, J. (2012). Annotated portfolios. *Interactions, 19*(4), 40–49. https://doi.org/10.1145/2212877.2212889
Keegan, M. (2018, April). *Hong Kong's 'cardboard grannies': The elderly box collectors living in poverty*. The Guardian. Retrieved January 1, 2024 from https://www.theguardian.com/cities/2018/apr/24/hong-kong-cardboard-grannies-elderly-box-collectors-recycling-poverty

Chapter 6
Noticing the Rhythms: Women Street Vendor Regulation and Resistance in the Historical Center of Guatemala City

Gustavo Fernandez, Andrea Echeverria, Katreen Boustani, Lucy Lemus, and Christena Nippert-Eng

Abstract We present a collection of images taken in and around a prominent, bustling plaza in Guatemala City's Historical Center. The plaza is a locus of social, political, and economic life in the city. As such, it is a fruitful site to notice the rhythmic processes by which public space continuously gets constructed and regulated in the everyday. The images, taken over a nine-month period, bring attention to the contestant, power-laden, and ever-shifting nature of the plaza. We fix our gaze on women street vendors in particular to visibilize their unique but often overlooked perspective.

Vendors form a crucial backbone of the plaza's social landscape. Yet they are also treated by the municipality as threats to the order of the space. We can read from the photos how this tension manifests in day-to-day socio-spatial rhythms, whereby street vendors work in varying states of mobility depending on the legality of their business, constructing technologies that aid them in their practice.

G. Fernandez · A. Echeverria
Guatemala City, Guatemala

K. Boustani (✉)
Syracuse University, Syracuse, NY, USA
e-mail: kboustan@syr.edu

L. Lemus
Queens, New York City, NY, USA

C. Nippert-Eng
Indiana University Bloomington, Bloomington, USA
e-mail: cnippert@iu.edu

S. An Blevis et al. (eds.), *Seeing Interaction Design*, Design Research Foundations,
https://doi.org/10.1007/978-3-032-14200-9_6

6.1 Introduction

Guatemala City, where it sits today, was first settled in 1776 after the capital city was transferred from the Panchoy Valley of present-day Antigua to the Ermita Valley (Aroche, 2023). The city limits remained largely unchanged for roughly a century before expansion accelerated. Now, they form the boundaries of an area referred to as *El Centro Histórico*, the Historical Center. At the heart of El Centro Histórico, situated within a checkerboard-like grid system of streets, one finds the main plaza of Guatemala City, nearly as old as the city itself. To its north stands the National Palace, renamed the Palace of Culture upon its conversion into a museum and tourist destination in 1943 (Garzano, 2014). To the east lies the Metropolitan Cathedral of Santiago of Guatemala. To the south, located along historic Merchant Street, is the Portal of Commerce—a longstanding hub of economic activity in the city, now converted into a shopping center. To the west runs *La Sexta Avenida*, or 6th Avenue, recently rebranded "Paseo La Sexta" as part of a municipal government-led modernization initiative to attract investment and tourism to the area.

Iterations of this urban form—a grid-plan town with a centrally located plaza—have appeared across time all over the world. Plazas are often crucial fixtures in the social, political, and economic life of a town or city. They host public gatherings, festivals, markets, and celebrations. They are sites of struggle and possibility, where peoples come together to voice collective grievances and organize themselves to transform their material conditions and reimagine social life. They reflect norms, practices, and values of social life. And, importantly, they reveal a great deal about the operation, organization, and reproduction of power in space. The plaza design in Guatemala City's Centro Histórico combines the grid-plan plaza form with a specific institutional arrangement—a place of worship, a government or military edifice (in this case, a palace), and a commercial center surround the plaza (Gómez, 2021). It shares this form with numerous cities and towns constructed throughout Latin America during the Spanish colonial period. This particular grid-plan plaza form, Setha Low argues, emerged as a product of the colonial encounter through processes of violence, erasure, resistance, adaptation, adoption, cooperation, appropriation, and negotiation. The form evolved throughout the sixteenth and early-seventeenth centuries in a series of urban design experiments as this encounter unfolded. It incorporates elements found in Spanish cities, such as the thirteenth and early-fourteenth century bastides and the Islamic Garden layouts, but also elements particular to indigenous Latin American architectural traditions, such as the grid pattern layout and cardinally oriented streets characteristic of Mexica urban planning. Certain acts of legislation, such as the Laws of the Indies—a series of policies enacted by the Spanish crown and enforced upon the "New World" in an effort to control the relationship between Spanish settlers and indigenous peoples—standardized this new urban form by codifying city planning and construction processes in areas under Spanish control, leading to its uniform replication in various regions throughout Latin America, as well as in Spain (Low, 2000). The plaza in Guatemala City was originally designed in 1778 by architect Marcus Ibáñez according to the ordinances passed by Spanish King Carlos III. It too followed norms set by some of the Laws of the Indies (Gómez, 2021).

The plaza in Centro Histórico has undergone countless changes in the centuries following its construction, looking starkly different now from its original design. Yet, by highlighting how these colonial era grid-plan plazas retain architectural, spatial, and physical elements from Spanish as well as indigenous Latin American traditions, Low figures them as "contested terrain[s] of architectural representation," illustrating how we can read the built environment as an embodiment of various cultural and political meanings of the past, but also, importantly, as an expression of power struggles and the tensions they produce—such as that between conquest and resistance (Low, 2000, p. 123). In this pictorial, we read the plaza of Centro Histórico as a contested terrain in more ways than one.

The name of the plaza provides an example. Since its original construction, the plaza has gone by different names: Plaza de Armas, Plaza Mayor, and now, officially, Plaza de la Constitución (Gómez, 2021). However, following a horrific fire that took place in the "Virgen de la Asunción" Home, a state center for girls some miles outside the city in March 2017, a movement was born to popularly rename the plaza *Plaza de las Niñas* in honor of the 41 victims, all girls, whose deaths people hold the state directly responsible for. The night before the fire, the girls held a demonstration, demanding to be released from the center on the grounds that they were being exploited and held in harsh conditions. They were locked up as punishment for this act of disobedience. The following day, as the fire raged on, police, guards, and local staff stood outside. Nothing was done to interfere in their deaths, and only 15 survivors escaped. The name *Plaza de las Niñas* is overtly political, meant not only to remember the girls who died, but to confront the hegemonic view of who the space of the plaza is meant for, and who is excluded from that vision. In addition to the name change, an altar was placed in the plaza as a symbol of denunciation against this act of state violence. It has been destroyed several times, but people keep coming to replace it.

Public spaces like the plaza are not static. And although often shaped by the strong influence of powerful elites, they are always co-constituted by many actors. As Doreen Massey writes, space is "the product of interrelation; as constituted through interactions, from the immensity of the global to the intimately tiny" (Massey, 2005, p. 9). It is "the sphere of the possibility of the existence of multiplicity in the sense of contemporaneous plurality as the sphere in which distinct trajectories coexist; as the sphere therefore of coexisting heterogeneity" (9).

Space is always under construction, always in the process of being made, never finished, and never closed (though some may try with impressive strength to close it). It is, as Lefebvre famously stated, *produced*—continuously so (Lefebvre, 1991). Though it is always ongoing, the production of space does not occur at a uniform tempo. Rather, it occurs at multiple overlapping timescales. Some features of a space last for fleeting moments and others, carved in stone, make their mark for many decades or centuries. In cases of infrastructural haunting, the remnants of long-gone designs continue to exert material force that shape a space as new designs overlay them, retaining aspects of their original form. Thus, we follow Lefebvre's assertion that space and time need to be thought together, and time's relation to the production of space understood nonlinearly.

To think about how everyday actions and interactions—the way someone moves from point to point across a lawn, the conversation between two people standing by a fountain, a police officer ordering a street vendor to pack up her wares and move—operate with and through a multiplicity of forces at different timescales to become long-lasting, regular features of certain spaces, but which remain subject to change forever at a given moment, we find it useful to think with *rhythms*. Lefebvre writes that 'everywhere where there is interaction between a place, a time, and an expenditure of energy, there is rhythm' (Lefebvre, Rhythmanalysis: Space, Time and Everyday Life, 2004, p. 15).

Rhythms are repetitions of movements and actions—sometimes linear, sometimes cyclical—which have phases of growth and decline (Edensor, 2010, p. 3). Analyzing socio-spatial rhythms gives us an eye for "how power is instantiated in unreflexive, normative practices but also side-stepped, resisted and supplemented by other dimensions of everyday experience" (Edensor, 2010, p. 2).

The images in this chapter were taken as part of our ongoing research project on women street vendors in Guatemala City, many of whom frequent the plaza. The broader focus of our work is on the personalized socio-technical systems these vendors construct to manage the boundaries between their home and work lives. The photos showcase the plaza from different angles at various dates and times over the nine-month span we conducted fieldwork. Read together, they highlight the ever-shifting nature of the plaza, and in turn draw attention to what remains constant amidst the uneven flows of people and objects—namely, the presence of street vendors. Vendors—their bodies, carts, goods, relationships, and selling practices—form a crucial backbone of the plaza's social landscape. Yet they are also treated by the municipality as threats to the order of the space, making them targets of legislation enforced by municipal police officers. We can read from the photos how this tension manifests in day-to-day socio-spatial rhythms, whereby street vendors work in varying states of mobility depending on the legality of their business, constructing technologies that aid them in their practice.

In October of 2023, our project took an unforeseen turn when protests and strikes sparked by outrage over government corruption erupted across the nation. To honor this important political moment, in some of the final images we reflect on protests as rhythms that shape the plaza and the mechanisms by which its order gets preserved.

6.2 Contributions

We share glimpses of the plaza from different angles at various dates and times to emphasize the contestant, power-laden, and ever-shifting nature of public space. We bring attention to the everyday interactions that produce and regulate such spaces through rhythms. We fix our gaze on women street vendors to visibilize their unique

Image 6.1 On the Corner of 8th Street and 6th Avenue, Sale of Goods Prohibited

but often overlooked perspective. In some of the final images we reflect on protests as rhythms that shape the plaza and the mechanisms by which its order gets preserved.

Image 6.1: At the Corner of 8th Street and 6th Avenue, Sale of Goods Prohibited.

To the right of a stone column, a tiled sign on the building's edifice reads, "Calle de Mercaderes" or, in English, *Merchant Street*. Outside a storefront with backpacks and purses on display, a municipal police officer stands, thumbs tucked into pockets. He looks into the camera, presumably observing the photographer. A young child grasps the hand of an adult and looks at the backpack and purse window display, pointing to a product on the other side of the glass window. 6th Avenue is crowded, teeming with life. People walk in the street and on the sidewalk. Some can be seen entering or exiting the many brick and mortar shops that line the street. A bright green sign fixed to a light pole reads, "Prohibidas las ventas en el espacio público" or, in English, t*he sale of goods is prohibited in the public space.*

Near the bottom of the sign, a piece of legislation is cited: "Artículo 9: Reglamento del Paseo de la Sexta" or, in English, *Article 9: Regulation of Sixth [Avenue] Walk*. The name Muni Guate is displayed at the top of the sign, shorthand for the municipality of Guatemala City.

The tension here—between the evidently booming economic activity on 6th Avenue and Merchant Street, and the ban on the sale of public goods—prompts a question: what kinds of sales are permissible, even encouraged, and what kinds are prohibited?

Image 6.2: At the Corner of 7th Avenue and 6th Street, Enforcers.

A man in a yellow button down and white trousers approaches the corner, gazing at the ground as if lost in thought. The difference in his position between the first and second photos allows us to sense of the passage of time. Two municipal police officers in motorcycle helmets patrol the corner. In the first photo, an officer appears

Image 6.2 At the Corner of 7th Avenue and 6th Street, Enforcers

to be observing the street vendor's cart sitting underneath a traffic light, but the helmet and dark glasses covering his face and eyes make it impossible to tell exactly where he is looking. Street vending in and around the plaza is heavily policed.

Gates, covered in green canvas signs, line the north and east edges of the plaza, blocking public entry. The signs advertise a cultural festival on Paseo de la Sexta. In the second photo, the bottom right corner of the green sign in prominent view reads "la ciudad es tu casa" or, in English, *the city is your home.* The statement prompts us to wonder, whose home is the city really, if certain areas are only accessible for some and not all?

Image 6.3: On 8th Street East of the Plaza, El Turno Malo.

A woman vendor packs up the items she has laid out for sale on a small blanket. She fits them into her backpack with care. She has been spotted by a municipal police officer on patrol, who stares at her while she packs. In a conversation with the woman, she tells us she knows the schedule of which officers work what days. Some are stricter about enforcing the rules prohibiting vending, others less so. The good days she calls "el turno bueno," *the good shift,* when she doesn't have to pack up or move. The bad days she calls "el turno malo," *the bad shift*. Today is the turno malo. Maybe more than any other image in the pictorial, we think this one captures what we mean when we refer to public spaces as constructed through socio-spatial

Image 6.3 On 8th Street East of the Plaza, El Turno Malo

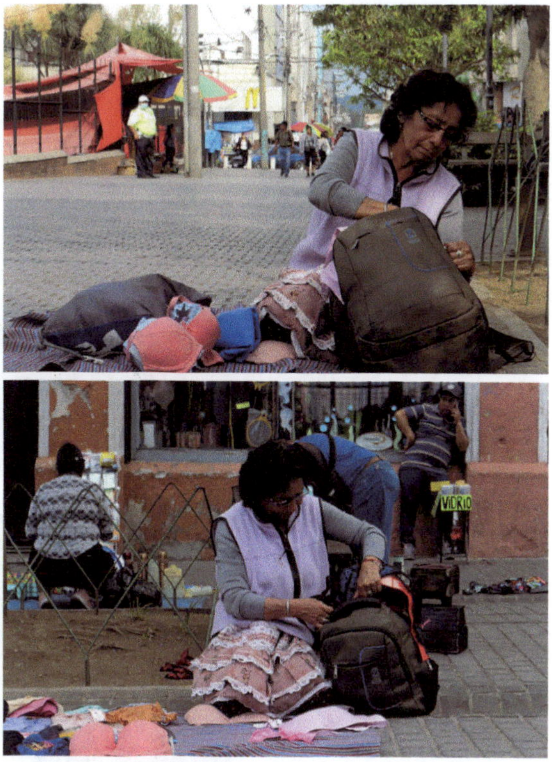

rhythms. Through the everyday back and forth, the good shifts and the bad shifts, the movements and adaptations, the space is created—serving certain purposes, fulfilling certain norms, and operating as a contested terrain of who and what belongs.

Image 6.4: Near the Fountain Inside the Plaza, Bubbles for Sale.

A woman in a blue cap blows bubbles with one of the many bubble wands she has available for sale. With her right hand, she dips the wand into the soapy bubble liquid. In her left hand, she holds multiple items—the liquid container, a dinosaur toy, and what appears to be a plastic bag. Her cart has four wheels making it easy to move around. Every inch of space is utilized to display the goods—brightly colored toys wrapped in plastic hang from white metal grids that line the cart on all sides. The bubbles are displayed prominently, sticking out from the top of the cart and tied tightly with string or maybe rubber bands. A young girl chases after some bubbles floating in the breeze.

Behind them, people are seated around the fountain talking, gazing around, eating corn, or checking a phone. Other vendors with their carts are scattered around the fountain. One cart on the opposite side appears to be selling similar colorful bubble wands, along with Guatemalan flags. The Cathedral stands tall in the background.

Image 6.5: Inside the Plaza, Vendors at Rest (1/2), Vendor in Motion (2/2).

Image 6.4 Near the Fountain Inside the Plaza, Bubbles for Sale

Image 6.5 Inside the Plaza, Vendors at Rest (1/2), Vendor in Motion (2/2)

In the top photo, two women vendors, an older boy, and two children sit chatting on a long stone park bench built into the ground of the plaza. One woman holds the child, her hands clasped around her chest. The other woman looks familiar—the same one selling bubbles in Image 6.4. She takes a bite of something while she sits. Three bubble wands sit on her waist. A dog lays down next to them as they talk, seemingly immersed in their conversation. The women seem to know each other. We believe they are family. On the right side of the photo is a vending cart that looks to be made from a shopping cart. Shirts, skirts, and dresses have been placed on hangers and hung in rows off the metal grids that line the cart on all sides. Two headless plastic mannequins wearing brightly colored shirts jut out from the top of the cart, only visible from the chest up. Another mannequin is attached to the side of the cart, fully visible in an orange, long-sleeve dress that hugs its figure. In the background of the photo, Paseo de la Sexta is visible.

In the bottom photo, the vendor previously holding the child walks the clothing-adorned cart back toward the center of the plaza, guiding it with her left hand. It looks heavy, but she moves it with ease. In the plaza's center, we see two more bubble wand and toy carts like those from Image 6.4, along with other vending stands whose wares are not clearly visible. The Palace can be seen in the background. A notable thing about this photo is something that can't be seen—the little girl has jumped inside the cart and is riding as the vendor pulls her! She is fully hidden from view.

Image 6.6: Inside the Plaza, Mobility (1/2), Fixity (2/2).

In the top photo, we see three vending carts, each with similar multicolored umbrellas shading them from above. It is nighttime and the lights hitting the front of the church in the background make it glow yellow like a full moon. The cart on the left is labeled with the Coca-Cola logo, appearing to sell bottles of water and a red sugary drink. The cart to the left sells phone cases, sealed in individually wrapped plastic and hanging from the cart except for the top row of products, which are packaged in boxes and tied to the cart. In the center of the photo, five people surround an Elotes Locos, or *crazy corns*, truck. The corn sits in a large, round metal dish. They are wrapped in plastic to keep them warm. A pair of tongs lays on top of them. An LED light has been fixed to the cart to help the vendor see in the dark. A metal grid basket has been fixed to the front of the stand. It is full to the brim with single serve chips, some which hang off the front of the basket. A bottle of hot sauce and a bottle of Tajin sit in the basket, as well. We presume these are to dress the crazy corn—yum! It is striking that all three carts appear fully mobile, having been altered with technical adjustments to ensure everything can be carted away easily at a moment's notice. Only the vendors' stools remain untethered to the carts.

In contrast, in the bottom photo, the vendor's wares are splayed out over a long table. She sits on one of three plastic stools, and two long yellow benches rest behind her. To her left, someone, maybe a customer or a thirsty family member, sits on a yellow bench about to take a sip from a cup. It's clear this vendor is in no hurry to leave and can take some time to pack up at the end of the night. We ascertain the photos represent vendors with different permit statuses who have adapted their selling spaces accordingly.

Image 6.7: Paseo de La Sexta, Policed (1/2), In Protest (2/2).

Image 6.6 Inside the Plaza, Mobility (1/2), Fixity (2/2)

In the top photo, taken March 5, 2023, in the late afternoon, two municipal police officers stand watch as crowds stroll along Paseo de la Sexta (6th Avenue). A skateboarder rides by them, ignoring their gaze. The road is blocked off from car traffic so pedestrians may walk freely. To the left of the road stand tall, smooth stone arches. To the right is the 6th Avenue entrance to the plaza. On days when gates line the plaza from most sides, closing it off, this is the only way to access the plaza. In front of the entrance, a woman takes down her vending stall. The two mannequins she had on display lay down by her side. They are dressed in undergarments, presumably what she was selling.

In the first week of October 2023 protests and strikes erupted across Guatemala, sparked by outrage over government corruption. The bottom photo was taken October 20, 2023, in the late afternoon. It is Revolution Day, which commemorates the popular uprising that overthrew dictator Jorge Ubico in 1944. By this time, the nationwide protests that began in early October are in full swing. The crowds are greater in number today. Most people seem to be heading north towards the large gathering of protestors. A mobile cart selling "ricas granizadas," *sweet slushies*, moves along the path in the same direction. A man standing above the crowds holds a microphone and speaks to the people. Many sit along the bottom of the stone arches, listening and participating in the demonstration. Street vendors line the right side of the road

Image 6.7 Paseo de La Sexta, Policed (1/2), In Protest (2/2)

selling candies, plastic horns, and flags tied to streetlamps and trees. The flags, many with the blue-and-white Guatemalan national design, billow in the wind. Municipal police are noticeably absent from this photo.

Image 6.8: Palacio Nacional de la Cultura, Surveillance.

Image 6.9: At the Corner of 8th Street and 6th Avenue, Gates.

One crucial role of public space is to "communicate civic sentiments and social resistance through its design and commodification. Public space is often about public protest" (Low, 2000, p. 183). This is reflected not only when spaces like the plaza become sites for protestors to convene and express discontent, or get appropriated and transformed through occupations, but also in the way responses to such acts by state actors become encoded into the spaces themselves. Sometimes, in the event an act of protest is "too successful" in threatening the state, "the public space is closed … gated and policed" (p. 184). Other times, it is "briefly closed down and redesigned in such a way as to discourage its continued use by the groups that are deemed undesirable, and then it is policed when it is reopened" (p. 184).

In the first photo, taken December 18, 2023, a surveillance camera can be seen hanging from the National Palace of Culture building overlooking the plaza. The camera was installed in 2015 following a series of anti-corruption protests which led to the resignation and arrest of former president Otto Pérez Molina. Now, eight years later, waves of anti-corruption protests return to the plaza. Though the target of today's political unrest is different from 2015, we can read the camera as an object

Image 6.8 Palacio Nacional de la Cultura, Surveillance

of continuity. It connects past and present cycles of social upheaval, representing the ongoing escalation of state surveillance.

At the corner of Merchant Street and 6th Avenue, the holiday season brings new festivities to the plaza and new gates that restrict the flow of foot traffic. Three municipal police officers are present. Each has their eyes locked on a man and young child, possibly father and son. The pair avert their gaze and move to pass through the gates. Do these two belong in the plaza? Will they belong tomorrow?

Image 6.10: Map of the Plaza.

To give readers a more comprehensive picture of the plaza's geography, we present some of the pictorial images as points on a map pulled from OpenStreetMap.

Image 6.9 At the Corner of 8th Street and 6th Avenue, Gates

6.3 Summary

In this chapter we presented a collection of images taken in and around a prominent, bustling plaza in Guatemala City's Historical Center. The images and their descriptions highlight the various rhythms that construct the plaza, from the turno bueno and turno malo (Image 6.3), to the periods of rest and movement that make up a woman vendor's day with her family (Image 6.5), to the rhythms of vending and policing during peace and during protest (Images 6.1, 6.2, 6.3, 6.7, 6.9). We show how the rhythms of moving and selling in the plaza are different depending on one's relationship to the municipality (Image 6.6). Lastly, we present a map view of the plaza and locate various images on this map to help readers situate the plaza in a different way.

Image 6.10 Map of the Plaza

Acknowledgements We give many thanks to the contributing editors for encouraging us to make this chapter. A special thanks to contributing editor Eli Blevis, who kindly met with us to share advice on taking good street photos and prepared us some delicious coffee. Our biggest thanks goes to Oscar Andrés Lemus, who now lives among the stars and in our hearts. Oscar planted all the seeds of this project; it has been a great joy to help them grow in his honor.

References

Aroche, K. (2023, January 3). *Traslado de la Ciudad de Guatemala al Valle de la Ermita.* Retrieved from Aprende en Guatemala: https://aprende.guatemala.com/historia/epocas-historicas/traslado-ciudad-de-guatemala-valle-de-la-ermita/

Edensor, T. (2010). *Geographies of Rhythm: Nature, Place.* Ashgate.

Garzano, C. (2014, August 23). *De Norte a Sur y de Sur a Norte.* Retrieved from Guate Historia: https://guatehistoria.com/de-norte-a-sur-y-de-sur-a-norte/

Gómez, A. C. (2021). Historicidad Urbana de la Plaza Mayor de la Constitución de la Ciudad de Guatemala. *Revista Avance*, 97–109.

Lefebvre, H. (1991). *The Production of Space.* Basil Blackwell.

Lefebvre, H. (2004). *Rhythmanalysis: Space, Time and Everyday Life.* Continuum.

Low, S. M. (2000). *On the Plaza: The Politics of Public Space and Culture.* University of Texas Press.

Massey, D. (2005). *For Space.* SAGE Publications.

Chapter 7
Film in More-Than-Human Design: Using Film to Envision Beyond the Human

Heidi Biggs, Doenja Oogjes, and Shaowen Bardzell

Abstract Posthuman theory asks how one might conceive human 'beingness' in more relational, post-anthropocentric ways. Based on this, recent work in HCI and design is exploring how to decenter the human in design, and methods and design processes have sprung up to imagine ways of seeing people as entangled in assemblages of non-human others. In this chapter, we are inspired by Gilles Deleuze who sees film as a technology which decenters the human eye and can be used for more-than-human ways of seeing. We created an annotated portfolio of our own videos, which we analyze and annotate to show how they illustrate posthuman concepts. Reflecting on our videos, we offer the concept using video to develop a posthuman gaze and offer ways film can be crafted and taken up in posthuman design practice.

7.1 Introduction

In recent years, some interaction designers have shifted away from a human- or user-centric approach to design and are exploring design through a more-than-human (MTH) lens. This MTH approach acknowledges how humans, their technologies, and other non-human entities (like plants, animals, or materials) are not discrete entities—user and used or object and subject—but might be more accurately described as *entangled*. Interaction designers are actively seeking ways to conceptualize, communicate, and design for these entanglements. One such way might be through film. Film, following media theorists and posthuman philosophers, is always already inherently a more-than-human way of seeing the world. The camera, the materiality of film, and film editing techniques move vision away from the human and offer many new perspectives and possibilities. Therefore, in the following, we use videos from our own MTH design practices, to reflect on how film and video can be used to

H. Biggs (✉) · S. Bardzell
Georgia Institute of Technology, Atlanta, GA, USA
e-mail: hbiggs7@gatech.edu

D. Oogjes
Eindhoven University of Technology, Eindhoven, The Netherlands

S. An Blevis et al. (eds.), *Seeing Interaction Design*, Design Research Foundations,
https://doi.org/10.1007/978-3-032-14200-9_7

see from more-than-human perspectives and be a tool for design researchers interested in the more-than-human design. We use sketches and annotation of video stills to illustrate the more-than-human perspectives and concepts they bring to life and then reflect on how visual media like film can be used in more-than-human design research. As tools for mediating perception and representing what is seen from MTH perspectives, we believe film and video offer world building and critical potential for more-than-human design practice.

The more-than-human (MTH) turn in interaction design is sometimes also referred to as nonanthropocentric design (Disalvo & Lukens, 2011), posthuman design (Forlano, 2017), or more-than-human design (Wakkary, 2021). MTH design orientations draw from a group of theorists like critical posthuman and process-philosophers Gilles Deleuze and Felix Guattari, Rosi Braidotti, and Karen Barad, as well as new materialists like Anna Tsing and Jane Bennett, and also science and technology scholars like Donna Haraway and Bruno Latour. Collectively, these theorists suggest that social forces and material processes, entangled into configurations, systems, and assemblages that 'become' together toward an action or a goal. These theories challenge the prevailing idea that the human subject is a discrete, rational, and/or objective individual (subject) who is hierarchically superior to the inert, non-agential, non-humans (objects) around them. In the posthuman/more-than-human turn, suddenly humans are vulnerable to and entangled within the forces of a material world which they are existentially enmeshed within.

These theories and their uptake in design research respond to complex problems like climate change and the rise of AI—where it seems humans can no longer think in terms of linear, hierarchical relationships to the natural world or see technology as discrete from their human users. Instead, it might be more accurate to see people as co-constructed and caught up in dynamic, relational systems with high stakes. Adding to recent interaction design research in this MTH agenda, we turn to film theory and film as a medium by suggesting film/video can be used toward building a more-than-human design practice. We draw on new media scholars (Kember & Zylinska, 2012; McLuhan & Fiore, 1967; Navas, 2018; Zylinska, 2017) and the film theory of Gilles Deleuze (1989, 2003) to discuss in detail how film and new media—or in our case everyday phone videos used in or around our design processes—mediate and extends human consciousness in ways that allow designers to see and know differently along more-than-human lines.

7.2 Media, Film, and More-Than-Human Envisioning

Film and video have been used in interaction design for many purposes. For example, interaction designers are familiar with the role film plays in storytelling and concept development both in research and professional settings. Designers use video to narrate or tell stories, present speculative futures (Dunne & Raby, n.d.), and craft corporate design concept videos ("Microsoft: Productivity Future Vision," n.d.; "Project Glass: One Day… - YouTube," n.d.). Video and film techniques are also used methodologically, for example, in the creation of prototypes (Homaeian et al., 2021; Krause, 2018; Truong et al., 2006) and storyboards (Homaeian et al., 2021; Krause, 2018). In addition, film has been used to explore more-than-human concepts through *thing ethnographies* that explore the 'lives' of non-human objects (Giaccardi et al., 2016), nature-watch cameras that capture the activity of non-human neighbors in people's yards (Gaver et al., 2019), and to explore technological mediation and more-than-human agency (Oogjes & Wakkary, 2017). We were inspired to expand the applications of film in interaction design research within MTH design. We noticed how film can be used to see beyond human-centered perspective like telling stories or showcasing speculative visions, towards use cases that leverage film/video's inherent quality of mediating reality and bringing forward more-than-human points of view. We draw from posthuman and more-than-human media theory to think of film as a philosophical medium that innately offers and illustrates MTH perspectives and theory.

The multiple ways that film and photography mediate the world in more-than-human ways have been discussed at length by media theorists and philosophers of technology. The mediating impact of emerging media environments was notably theorized by McLuhan and Fiore who argued that the materiality of information (the medium of dissemination, the form it takes) shapes environments of thought and culture—arguing that computers and television were radically shifting and shaping thought, culture, and behavior (McLuhan & Fiore, 1967). More recently, new media theorists have begun to move beyond the interpretive meaning of film and photography to the more-than-human agency and mediating entanglements of photography and film, especially as they become increasingly automated (CCTV, Ring Cam, and the like) (Navas, 2018). For example, inspired by film theory of Gilles Deleuze, Media theorist Joanna Zelensky theorizes posthuman photography, both as it becomes more separate from human agency, as well as reflecting on the inherent non-human agency in photography. These theorists draw forth questions of what media sheds light on, how it is becoming increasingly divorced from human agency, and also how it mediates human-beingness. These are questions we draw forward as well, asking how film can be used in MTH design research, which is inherently about making, shifting, and intervening.

Ultimately however, we were inspired by Gilles Deleuze's film theory to think about how we use short videos to see in 'more-than-human' ways. Deleuze saw film as a technology that expands philosophical thinking through encounter with non-human ways of seeing (Colebrook, 2006). He suggests that humans need to confront what is not human to unlock the potential of philosophical thought by finding ways of seeing 'outside' of human experience. Deleuze suggests that through non-human technologies like cinema, we can confront, "the unthought, the accidental, and the unthinking" (Colebrook, 2006, 4) which expands philosophical thought by finding ways to see beyond the 'technology' and limits of the human body and human perception.

Deleuze explains how film is a unique more-than-human material as film extends beyond human perception in three ways. First, the cinematic image can be edited, enacting 'cuts' between multiple perspectives. These cuts give the impression that there are multiple ways to look at an object or scene at once. This technique illustrates the idea of multiple positionalities. Second, more-than-human or extra-human movement of the camera can move differently than the every-day human gaze. Camera movement adds to narrative through the speed and feeling of the camera as it moves through and 'sees' the world of the film. Third and finally, cinema can move through time in non-linear or non-human timeframes. For example, moving through montages, or showing flashbacks, or capturing a reaction in slow motion. Ultimately, Deleuze suggests film can pull humans through space and time in ways that expand beyond what humans can perceive with the imaginative lens of the camera. These are three high-level ways Deleuze looks at film—however, we also note that he dives deeply into the significance of much more than these in his two collected works on film theory.

In a small applied study, in the following we unpack the ways that video from our smartphones can extend our vision and capture moments of relationality and entanglement with non-human others. As designers, we extend the work of media theorists but putting video into the service of our design practice and design questions about how more-than-human ways of seeing influence our practice and process.

In our work, we explore everyday video we took with cameras on our smartphones. The videos were selected from a four-year range of photography 2017–2021 by Biggs and Oogjes, who are both engaged in MTH design practices. The videos either come directly from experiments with out more-than-human design research practice or in a few cases, are taken from moments caught in daily life by us as researchers sensitized by our more-than-human design practice. Therefore, the examples we provide are less cinematic and more like video-sketches. In each image, we include a sketch of how the video was made so that practitioners could see the 'making of' the image as well as still images from the videos themselves.

7.3 Contributions

In the following, we contribute a theoretical account of how film is inherently more-than-human and how the visual material of film, as such, can be leveraged into more-than-human interaction design practice. This builds on prior work in HCI in MTH research which deployed video, but opens avenues for expanded development of film/video as a MTH tool by connecting MTH design research with media theory. We ultimately reflect on how video is entangled with our own practices of being MTH design researchers and ask how video helped us develop and articulate MTH ways of seeing.

Image 7.1: Pinhole Gondola.

This video was taken in a gondola connecting two mountain tops. The phone camera was positioned on top of a small hole in the floor of the gondola (used to drain water from melting snow) through which the video was filmed. This set-up created a pinhole-type camera which created a tilt-shift effect which made the world seem tiny and strange. This video built a nonhuman perspective as it extended and

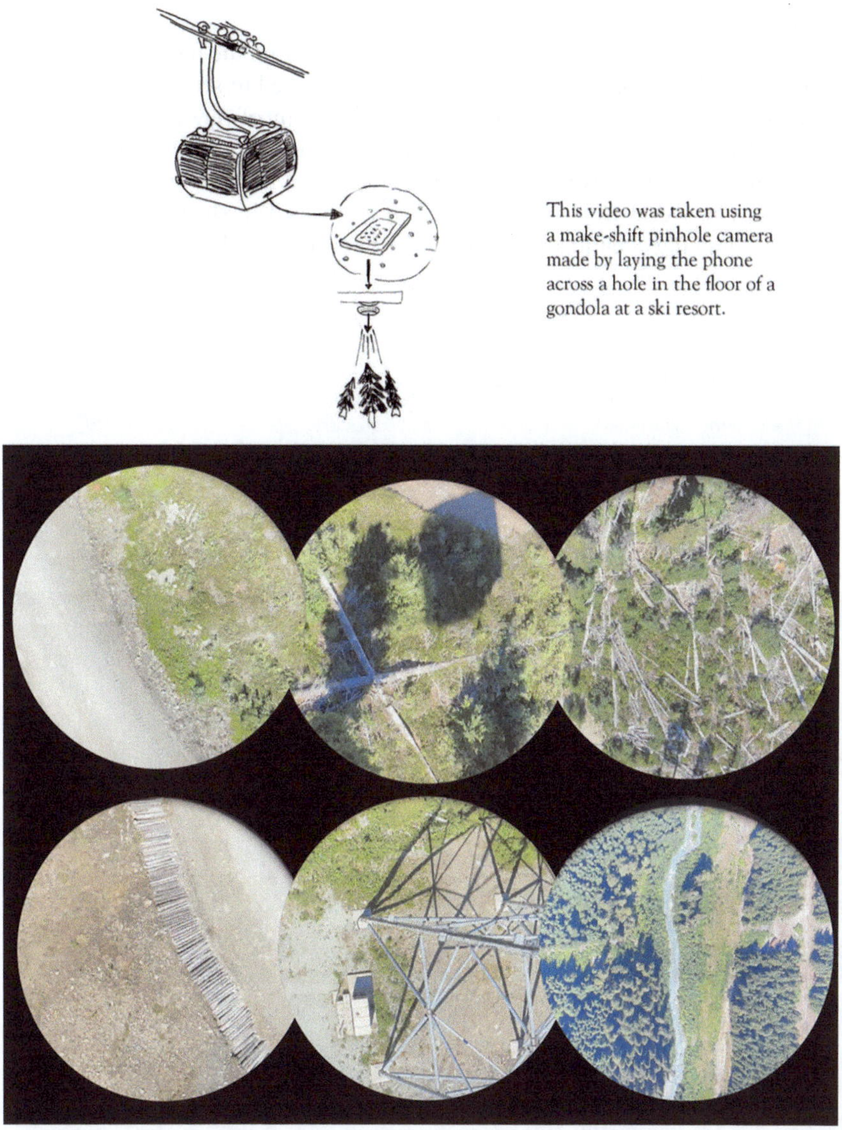

Image 7.1 Gondola pinhole

mediated human vision by going beyond what is visible with the human eye, taking a video from high above the ground, and creating miniaturizing effect on the images. The pinhole tilt-shift caused a slight defamiliarization of the world below as well, making it seem small, fragile, and precious. The sketch illustrates the phone was placed on the floor of the gondola and how the lens transomed the landscape slightly into a tilt-shift.

Image 7.2: Sock Scope.

Sock-scope shows still-frames from a video taken from the inside of a sock as it was mended. Reminiscent of a laparoscopic camera used to see inside a body, this video peers out from the inside of the sock, bringing a point of view that gazes upon the sock mender and slowly loses sight of them as the hole is closed. This helps the viewer gain a reflexive perspective on their own actions from the non-human perspective of a sock. The world from outside the sock is grasped in shadows, stitches, and fragments, drawing attention to the interiority of every-day objects, the pathways of

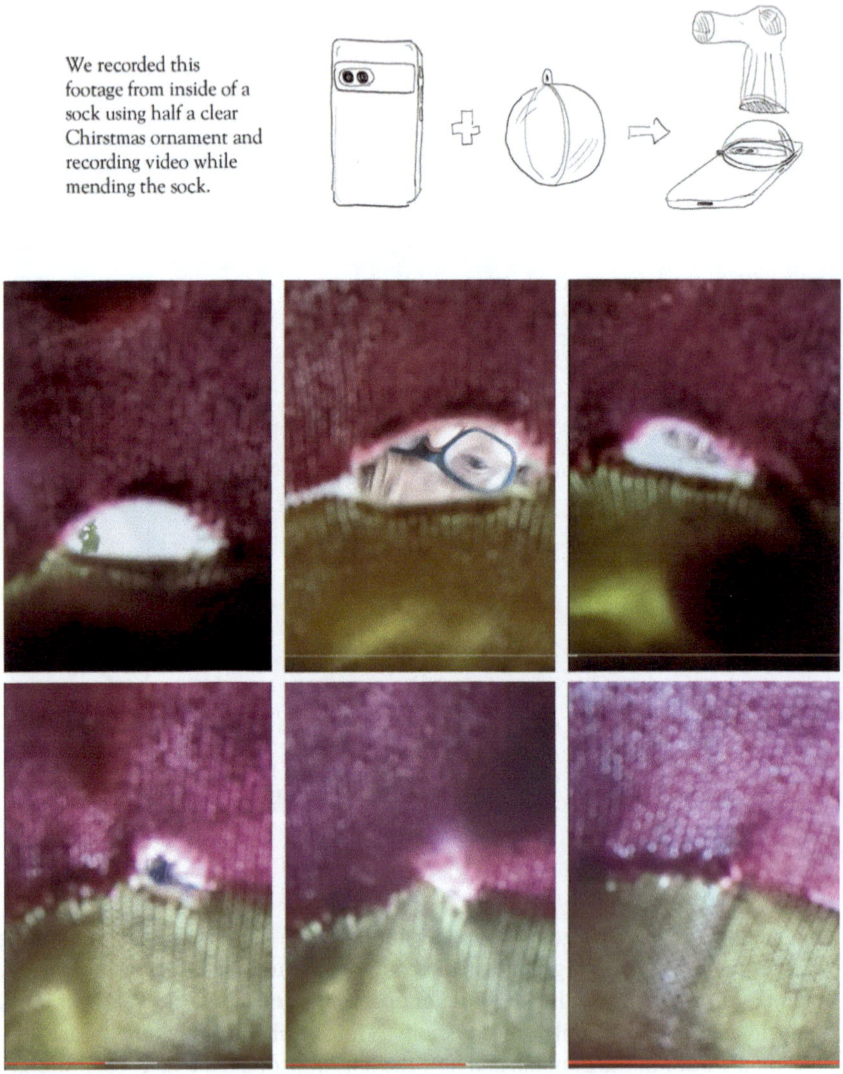

Image 7.2 Sock-scope

thread, the liveliness and lifecycle of garments, and the acts of care of mending. The sketch shows how the film was taken by using half of a clear Christmas ornament which was placed over the phone camera, then shoved into a sock, then mended on top of, to capture the interior view of a sock.

Image 7.3: Missed-takes.

This video cycles through sets of photos that were accidentally taken with a phone camera. Oogjes collected 120 images from herself and her friends over several years.

Image 7.3 Missed-takes

In the video, Oogjes brought together the collection in categories to show loose connections between common subjects. These categories are animals, lamps from below, floors from above, and blurred movement. As the video progresses, the photos shift and are displaced from their categories. Drawing from some of the new media theorists, these photos show the agency of a phone's camera removed from the desires of a human viewer.

Image 7.4: Bike/Flock/Phone.

Biggs took this video on their bike while riding through rural Mississippi exploring the non-human agency of the Mississippi River. Biggs noticed the birds in the tilled fields would periodically take to the sky and flock in a murmuration. They took this video by riding with one hand on the handlebars, one hand holding the phone up and following the motion of the birds. The result is a looping, swooping, video where multiple directions of human and non-human movement coalesce in a dizzying and beautiful video. This video represents three axis of relational movement balancing human and non-human orientations: the bike going forward, the birds swarming around, and Biggs' arm holding the phone, swinging around to catch the flock.

Image 7.5: Water Droplet Lens.

Oogjes was on a public transit train and noticed a raindrop on a train window, subtly enlarging the view behind it. The video is shot through this found raindrop lens and peaks into people's living spaces as the train traveled through the city. Oogjes captured a found assemblage with their camera of the movement of the sky train, the lens of the water droplet, and the passing apartments. The sketch shows the assemblage of how the video was captured using a found water droplet and asks us to explore how every-day natural occurrences can help us see differently, playfully unlock the imagination, and become a new lens onto the world.

Image 7.6: Wet Wash Wetlands.

This video was made during a painting process that imagined wetlands before and after settlement. It conceptually explored how geographies and map lines cannot be 'hard' or 'solid' on wetlands and watersheds through this material wet-wash process. The video was taken by cutting and editing a longer video taken using a tripod which suspended the phone over their painting (shown in the sketch). This video captured the way the paint spread into the texture of the paper and the colors blended into each other. Biggs used this video to think about how water systems and wetlands don't hold borders or boundaries of a map, and that polluted geographies which might be represented as stable on a map, will blur and blend in water systems and wetlands. The video's focus on the material process of painting captures the texture of the paper and the action of the paint over time. Film allows for a level of abstraction, and metaphor, where material processes become deeper meditations on non-human lifecycles of wetlands.

Image 7.4 Bike/flock/phone

7.4 Summary

In summary, we offer a visual analysis of the way that a visual medium (film) can be seen as a tool and collaborator in MTH design research and envisioning. We see each exploration as exploring various facets of MTH knowledge and different

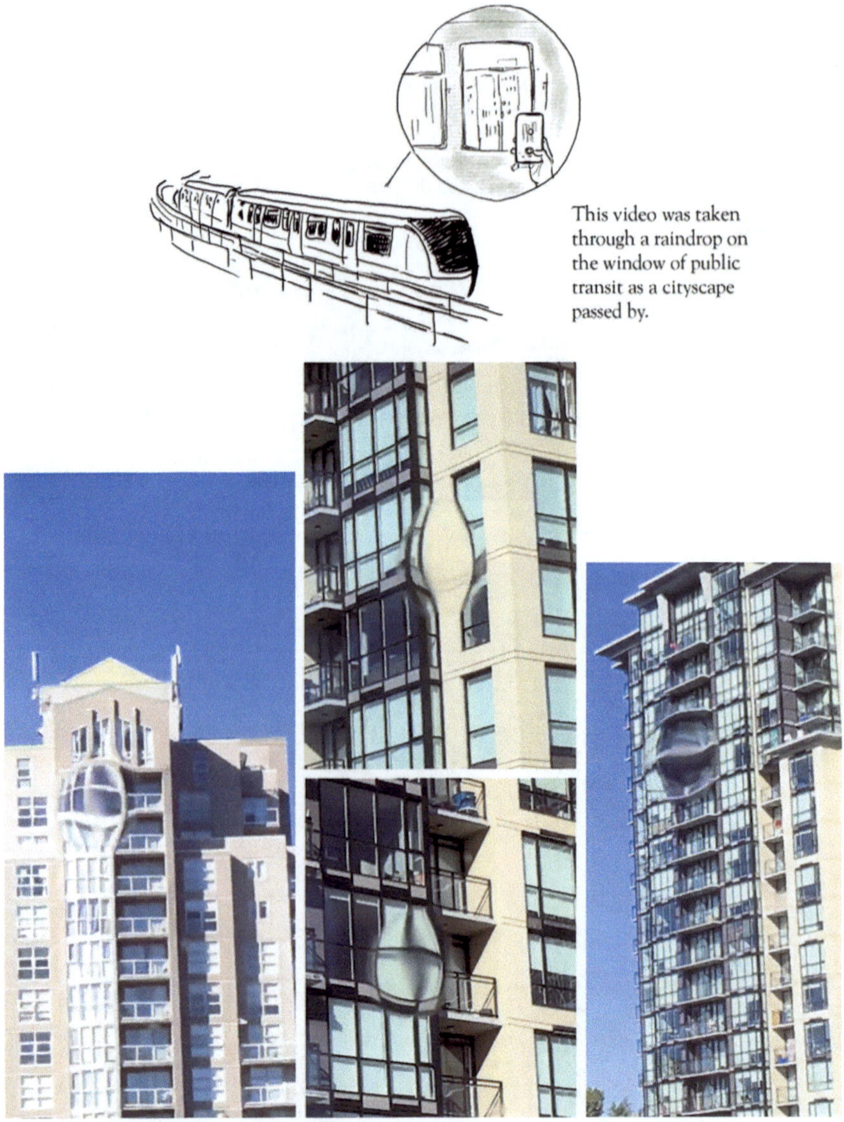

Image 7.5 Water droplet lens

design sensibilities. Some of the examples explore how the video camera can become entangled with everyday phenomenon to extend the eye and create delightful new perspectives on the world in terms of scale and visual quality. The gondola pinhole extended what could be seen by the naked eye and made the world suddenly miniature and fragile, while the water droplet lens extended the camera's lens, magnifying

Image 7.6 Wet wash wetlands

parts of the world going by. Some reflect on materiality, process, craft, and non-human perspectives together like the Sock-Scope and the Wet Wash Wetlands. Others reflect on non-human agency of the camera itself like Missed-Takes. Others represent experiences with non-human others in relational movement like bike/flock/phone. Each film combines the unique agency and potential of film with the materiality or experience of design practitioners. We suggest that by acknowledging surrounding media theory, we might begin to include cameras, film, and emerging media more purposefully in our more-than-human design practices, asking what we shed light on, and how the cameras in return mediate and show us the world in new ways.

References

Colebrook, C. (2006). *Deleuze: A Guide for the Perplexed*. Bloomsbury Publishing.

Deleuze, G. (1989). *Cinema 2: The Time Image*. University of Minnesota Press.

Deleuze, G. (2003). *Cinema 1: The Movement Image* (7th ed.). University of Minnesota Press.

Disalvo, C., & Lukens, J. (2011). Nonanthropocentrism and the Nonhuman in Design: Possibilities for Designing New Forms of Engagement with and through Technology. In *From Social Butterfly to Engaged Citizen*. MIT Press. https://doi.org/10.7551/mitpress/8744.003.0034.

Dunne, A., & Raby, F. (n.d.). Dunne & Raby Technological Dream Series: No. 1, Robots, 2007. Accessed February 11, 2023. http://dunneandraby.co.uk/content/projects/10/0

Forlano, L. (2017). Posthumanism and Design. *She Ji, 3*(1), 16–29. https://doi.org/10.1016/j.sheji.2017.08.001

Gaver, W., Boucher, A., Vanis, M., Sheen, A., Brown, D., Ovalle, L., Matsuda, N., Abbas-Nazari, A., & Phillips, R. (2019). My Naturewatch Camera Disseminating Practice Research with a Cheap and Easy DIY Design. In *Conference on Human Factors in Computing Systems—Proceedings* (pp. 1–13). https://doi.org/10.1145/3290605.3300532

Giaccardi, E., Cila, N., Speed, C., & Caldwell, M. (2016). Thing Ethnography: Doing Design Research with Non-Humans. In *DIS 2016—Proceedings of the 2016 ACM Conference on Designing Interactive Systems: Fuse* (pp. 377–387). https://doi.org/10.1145/2901790.2901905

Homaeian, L., Wallace, J. R., & Scott, S. D. (2021). Joint Action Storyboards. In *Proceedings of the ACM on Human-Computer Interaction* 5 (CSCW1). https://doi.org/10.1145/3449102

Kember, S., & Zylinska, J. (2012). *Life after New Media: Mediation as a Vital Process*. MIT Press.

Krause, R. (2018, July 15). Storyboards Help Visualize UX Ideas. Neilsen Norman Group. https://www.nngroup.com/articles/storyboards-visualize-ideas/.

McLuhan, M., & Fiore, Q. (1967). *The Medium Is the Massage*. Bantam Books.

"Microsoft: Productivity Future Vision." (n.d.). Accessed February 11, 2023. https://www.youtube.com/watch?v=w-tFdreZB94&t=12s

Navas, E. (2018). 10 the Assemblage Gaze of Media and Humans. In *Art, Media Design, and Psotproduction: Open Guidelines to Appropriation and Remix*. Taylor & Francis Group.

Oogjes, D., & Wakkary, R. (2017). Videos of Things: Speculating on, Anticipating and Synthesizing Technological Mediations. In *Conference on Human Factors in Computing Systems—Proceedings*, 2017-May (pp. 4489–4500). Association for Computing Machinery. https://doi.org/10.1145/3025453.3025748.

"Project Glass: One Day... - YouTube." (n.d.). Accessed February 11, 2023. https://www.youtube.com/watch?v=9c6W4CCU9M4.

Truong, K. N., Hayes, G. R., & Abowd, G. D. (2006). Storyboarding: An Empirical Determination of Best Practices and Effective Guidelines.

Wakkary, R. (2021). *Things We Could Design: For More Than Human-Centered Worlds*. MIT Press.

Zylinska, J. (2017). *Nonhuman Photography*. MIT Press.

Chapter 8
"You Should Get to Know Each Other": 10 Years of Sketching in HCI

Makayla Lewis and Miriam Sturdee

Abstract It is a visual tale of two sketchers: from meeting, to collaborating, to growing, and to considering the future of sketching, their work, and their connection. This chapter demonstrates the power of sketching and the vital role of empathy in successful sketch communication, research, and interaction with others. This chapter only suggests that some should work in such a fashion. It puts a lens on our story in the holes and denotes the possibility of creative expression in an overly digital age.

8.1 Introduction

Sketching is still relevant (Goldschmidt, 2017). Ten years ago, a mutual acquaintance suggested that we connect over sketching. A connection was formed using social media, ideas and practice were exchanged, and common ground was formed. Over the space of ten years, this loose connection has forged a partnership which not only solidifies the place of sketching within our research in Human–Computer Interaction but also has inspired further collaboration, exploration and innovation as an analogue space carved out into an increasingly digital world, sketching bridges a precarious gap between existing processes and change. By sharing our journey, we want to form new connections, explore the existing story through a new lens, and suggest future directions not only for Sketching in HCI but also for other collaborative endeavours.

M. Lewis (✉)
Kingston University, Kingston Upon Thames, UK
e-mail: m.m.lewis@kingston.ac.uk

M. Sturdee
University of St. Andrew's, St Andrews, UK
e-mail: ms535@st-andrews.ac.uk

S. An Blevis et al. (eds.), *Seeing Interaction Design*, Design Research Foundations,
https://doi.org/10.1007/978-3-032-14200-9_8

8.2 Contribution

We present a retrospective of ten years of sketching in HCI practice, research, and education, exploring the cross-disciplinary nature of sketching via a reflective journey from the point of view of independent practice through to a collaborative endeavour that has snowballed into a worldwide community and network of research. We pinpoint areas of good practice, share successes, and consider the potential for parallel journeys in other artistic mediums. Each set of images also corresponds to a 'take-away' message, included in the description, designed to highlight potential collaborative connections and ideas.

Image 8.1: Visual Tale of Two Sketchers.

This image captures the breadth of our collaborative work over the past 10 years since we connected on social media. It is a visual tally and celebration of the varied possibilities for collaborative work in HCI, seen through the lens of Sketching in HCI.

Message: Collaboration is a network of people, possibilities, venues, and outputs—do not limit yourselves to a single problem space or interaction.

Images 8.1 to 8.4: Connection.

Visual representations of our connection using a variety of imagery from 2018 to 2023 (Image 8.1). Image 8.2 (top) depicts a digital illustration from 2019, followed by (middle images) sketches from 2020 and 2022, and finally, a 'silly' activity for our 'Sketching in Human Computer Interaction: A Practical Guide to Sketching Theory and Application' book: 'sketch one another' non-dominant hand and looking away from the paper/canvas. Images 8.3 and 8.4 respond to a prompt, "What does our connection look like now?" (Lewis et al., 2018; Sturdee et al., 2022; Lewis et al., 2023; Lewis & Sturdee, 2024). Image 8.2 is a mixed media sketch depicting a visit to the Natural History Museum London to identify a mineral (Mica) that represents the strength, layered approach to sketching and our sharing to anyone interested in sketching in HCI. Image 8.4 echoes Image 8.3 through the 'digital glass' of sketch creativity, responding in kind to the inspiration and environment we find ourselves in on a day-to-day basis. One is urban area and surrounded by objects, people, and places. One rural area, considering the natural beauty and decay of green spaces. These images show the progression of our connection from being strangers to being professional to becoming playful and abstract.

Message: Creativity is a cornerstone of collaboration; non-academic interactions can inform ideation and form the bedrock of research outputs.

Image 8.5: Conversations.

Ninety per cent of our conversations occur on X (formally Twitter); we met on X and cannot let it go; it is our platform of choice; thousands of direct messages have been sent over the years—mixing professional and work life, conversations include art practice, research decision-making, rants, laughter, and excessive photographs of our pets. All our publications and event materials (papers, book chapters, journals, courses, special interest groups, etc., have begun in that little private corner of

Image 8.1 Visual tale of two sketchers. Pens and markers, Miriam Sturdee, 2023

X. Image 8.5 is a snapshot of our everyday communication: daily life, publication updates and decisions, screenshots of thoughts, a celebration of life events, Ren and Um (our pets), hobbies, and support in professional and personal nature. Conversations are born of respect, honesty and support, the key to friendship and research collaboration; often, researchers try to keep these separate. However, this relationship enriches both aspects of our lives.

Image 8.2 Connection No. 1. Through the years! Digital (Procreate App, pens and markers) portraits, Makayla Lewis and Miriam Sturdee, 2023

Message: Collaborative exchanges can also be social exchanges; a collaboration is also a research friendship founded on shared values and mutual support.

Images 8.6 and 8.7: Sharing the Bounty.

Sharing through publication for many is the most structured approach, e.g. Image 8.7 (Lengyel et al., 2023; Lewis & Sturdee, 2023; Sturdee et al., 2021; Sturdee et al., 2022). However, a more interactive approach is preferred, as it shows our sketch

Image 8.3 Connection No. 2. Pens, colour pencils, photograph and procreate App, Makayla Lewis, 2023

skills, connection, and creative environments (at home for remote sessions, 'show and tell' for in-person sessions, and a combination of both for hybrid sessions). The goal is to "bounce off" each other, demonstrating the friendly aspect of collaborative sketching, constructive critique, and the power of sketching in HCI. Image 8.8 is a

Image 8.4 Connection No. 3. Pens and markers, Miriam Sturdee, 2023

snapshot of our live sketching outcomes from meetups, events, conferences (Lewis & Sturdee, 2023), fireside chats, and in the classroom (lectures and workshops).

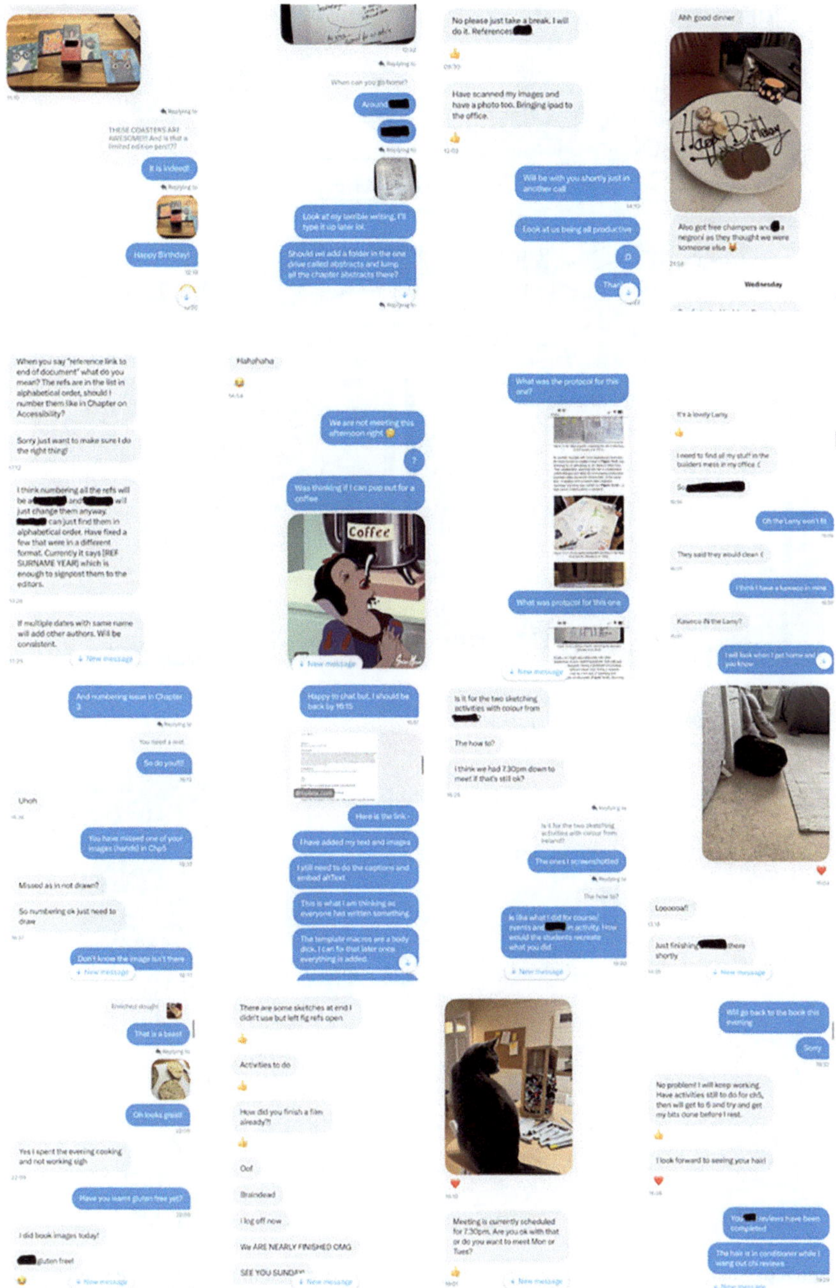

Image 8.5 Conversation—extracts of Direct Messages from X (formally Twitter). Screenshots from October and November. Makayla Lewis and Miriam Sturdee, 2023

Image 8.6 Share the Boundary No. 1. Pen and marker on Miro (online collaborative whiteboard). iPad Pro and Apple Pencil. Makayla Lewis and Miriam Sturdee, 2018 to 2023

Image 8.8: Spending a lot or too much time together.

The rise of remote (and hybrid) The rapid expansion of remote and hybrid working arrangements since 2020 has been a 'blessing' for our research, creating conditions for greater productivity and substantially expanding both the scope and volume of our outputs. (e.g. Images 8.7 and 8.8). The time afforded due to reduced commuting, digital multi-tasking in our everyday work practice, and the many Microsoft Teams calls, sometimes to co-work, prod away at unrelated tasks, or vent about daily life, has led to a blur between life and work, the phrase "whenever is good for you" has become commonplace. This blur is echoed in Image 8.4; this would be concerning (or surprising) for many, although it works for us because the amalgamation allows for easy jumping between both worlds: observing each other social interactions, workload, and personal issues whilst having a joint understanding of what needs to be done leads to our empathic and creative work environment. Image 8.9: we ensure communication, empathy, and flexibility are rooted in our connection, thus safeguarding effective professional working, personal social interaction, project management, and sketch creation. The incorporation of 'No More' by Ruff Endz lyrics, circa 1999,[1] is one of our favourite songs when co-working. We believe sketching is a powerful communication tool that represents the creator and the user in an empathic sense; these sketches are flexible in their creation, content, and format to ensure they suit the creator and reader. Without incorporating these skills into our daily life, our connection, research, and teaching would not be where it is today.

Message: Embrace familiarity and mutual support; co-working sessions do not necessarily mean 'meetings'; they can be shared safe spaces for research and ideation.

Image 8.9: The Future

The future is unknown, but we will continue communicating, exploring, and sharing as a sketching team. Image 8.9 depicts our communication, exploration, and the connected sketch possibilities.

Message: Successful collaborations do not have a defined ending; leave the door open for further possibilities.

8.3 Summary

This chapter tasks a visual approach to sharing the life and works of contemporary sketching by HCI researchers living in different areas of the country; it depicts our story, compassionate work practices, our need to share with the world, and the connection that has grown slowly over time.

A successful collaboration can be many things. A mutual understanding, a friendship, a drive to do better, share knowledge and best practices, explore new directions and the old, and bring others into orbit. Lupi and Posavec's (2016) 'Dear Data'

[1] Please listen to this song, where you view all the images in this chapter. We have chosen to provide a Spotify link as this is the platform where we listen to various 1990s RnB playlists: https://open.spotify.com/track/2WfhlEjoUII31H6imnQdvF?si=439d1d32a67d4af7

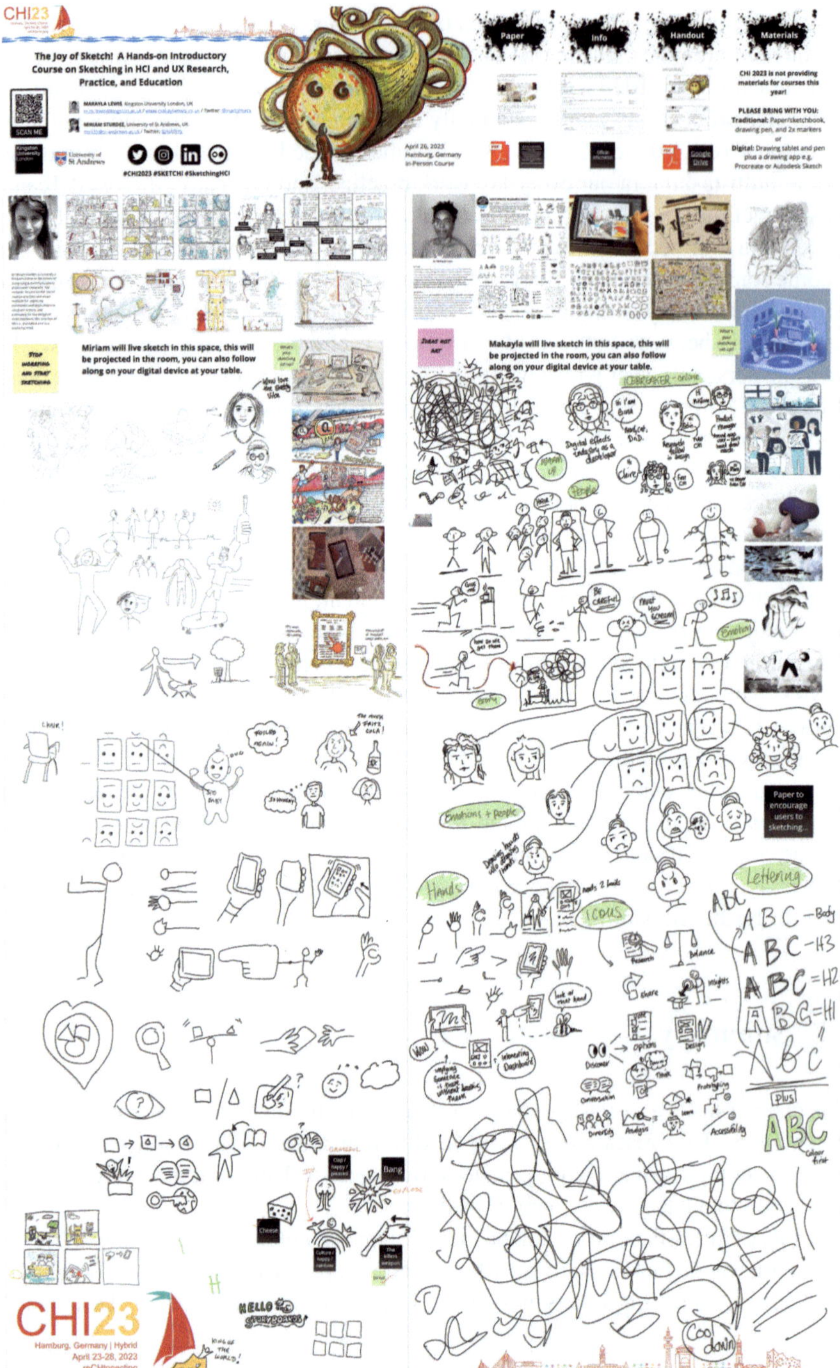

Image 8.7 Share the Boundary No. 2. Live sketch demonstration from ACM SIGCHI 2023 conference course 'The Joy of Sketch'. Makayla Lewis and Miriam Sturdee, 2023

Image 8.8 Spending a lot, or too much, time together. Microsoft Teams Screenshot and Procreate App. Makayla Lewis, 2023

bridged distance and data and showed mutual respect and playful creativity in practice. In 'Research Journeys', Sturdee et al. (2020) surface the practical aspects of conducting research in the field and within the academia, giving a behind-the-scenes glimpse into the world around us as we interrogate and explore.

Image 8.9 The future. Procreate App. Makayla Lewis, 2023 from our Sketching in Human Computer Interaction: A Practical Guide to Sketching Theory and Application (Lewis & Sturdee, 2024)

This pictorial chapter is a playful look into ten years of collaborative practice and research and a glimpse into the continued relationships that support our academic output. We share these personal, autoethnographic, mutual experiences as authors, academics, artists, and friends. Laid bare, collaboration is a collection of experiences, building a world Coulton et al. (2017), of which the majority is hidden from view, like an iceberg. However, research is the sum of many parts and people, all of which should be celebrated.

Sketching is best served when shared.

Acknowledgements We want to thank our colleagues, family and friends, students, and research participants for accompanying us on this journey of sketch discovery, exploration, and sharing.

References

Coulton, P., Lindley, J. G., Sturdee, M., & Stead, M. (2017). Design Fiction as World-building.

Goldschmidt, G. (2017). Manual Sketching: Why is it Still Relevant? In *The Active Image: Architecture and Engineering in the Age of Modeling* (pp. 77–97).

Lengyel, D., Kaipainen, K., Sturdee, M., Heron, M., , M., & Liddle, J. (2023). Hands-on Workshop on Tabletop Role-playing for Inclusive Design: Imagining Sustainable Futures for 'Older Adults'. In *Academic Mindtrek 2023 Conference.*

Lewis, M., & Sturdee, M. (2023). The Joy of Sketch: A Hands-on Introductory Course on Sketching in HCI and UX within Research, Practice, and Education. In *CHI Conference on Human Factors in Computing Systems Extended Abstracts (CHI '23 Extended Abstracts), April 23–28, 2023, Hamburg, Germany*. ACM.

Lewis, M., & Sturdee, M. (2024). *Sketching in Human-Computer Interaction.* Springer.

Lewis, M., Sturdee, M., Hoang, T., Fdili Alaoui, S., Strohmayer, A., & Samuelsson-Gamboa, M. (2023). Futuring (CHI)Art: Building a Collective Common Future. In *CHI Conference on Human Factors in Computing Systems Extended Abstracts (CHI '23 Extended Abstracts), April 23–28, 2023, Hamburg, Germany*. ACM.

Lewis, M., Sturdee, M., & Marquardt, N. (2018). Applied Sketching in HCI: Hands-on Course of Sketching Techniques. In *Extended Abstracts of the 2018 CHI Conference on Human Factors in Computing Systems* (pp. 1–4).

Lewis, M., Sturdee, M., Miers, J., Davis, J. U., & Hoang, T. (2022). Exploring AltNarrative in HCI Imagery and Comics. In *AltChi Paper to 2022 CHI Conference on Human Factors in Computing Systems. Association for Computing Machinery*, Article 164.

Lupi, G., & Posavec, S. (2016). *Dear Data.* Chronicle books.

Sturdee, M., Lewis, M., & Miers, J. (2022). Do Humans Dream of Digital Devices? Subconscious User Experiences and Narratives. In *Creativity and Cognition (C&C '22). Association for Computing Machinery.*

Sturdee, M., Lewis, M., Strohmayer, A., Spiel, K., Koulidou, N., Alaoui, S. F., & Urban Davis, J. U. (2021). A Plurality of Practices: Artistic Narratives in HCI Research. In *Creativity and Cognition (C&C '21). Association for Computing Machinery*, Article 35, 1.

Sturdee, M., Robinson, S., & Linehan, C. (2020). Research journeys: Making the invisible, visual. In *Proceedings of the 2020 ACM Designing Interactive Systems Conference* (pp. 2163–2175).

Part III
Pushing Boundaries

Chapter 9
Pushing Boundaries: Design for Respect, Qualities of Focus, Stillness and Motion

Shunying An Blevis and Eli Blevis

Abstract This chapter on pushing boundaries samples several papers and pictorials that push at the edges of what can be understood to be interaction design. The methods are photographic, not so much as Art—although the merits of the photographs as photographs matter here, not so much as Ethnography—because the photographs are not really data, but rather seeing interaction design in the sense of seeing the politics of interactivity through the lens of photography.

9.1 Contributions

We give examples of pushing boundaries in several contexts, from (a) the signs—*literally*—that call for Unity, Tolerance, Respect and Peace in a protest, to (b) label designs that invoke respect or disrespect for others, to (c) the abstraction of entirely defocused images that challenge notions of truth in representation, to (d) the physical movement of crowds, to (e) the marketing and consumption of personal digital technologies, the metaphorical movement of consumer crowds.

Image 9.1 Signs at a Rally in Support of Asians and Asian Americans and Pacific Islanders Communities, Shunying An Blevis and Eli Blevis, Photograph, 2023

In February of 2023, several community groups organized a rally at Indiana University's Bloomington campus in response to a racially motivated assault against

Photographs in this chapter are selected from *Stillness and Motion, Meaning and Form* (Blevis, 2014), *Qualities of Focus* (Blevis, 2017), *Seeing What Is and What Can Be: On Sustainability, Respect for Work, and Design for Respect* (Blevis, 2018), *Nonlinear Design Thinking: The Limitations of Awareness in the Search for Climate Care* (Blevis and Blevis, 2022), and *Advancing Non-Linear Design Thinking* (Blevis et al., 2023). The texts are adapted and summarized from these sources, with some additional interpretations.

S. A. Blevis · E. Blevis (✉)
Luddy School of Informatics, Computing, and Engineering, Indiana University, Bloomington, IN, USA
e-mail: eblevis@iu.edu

S. A. Blevis
e-mail: san@iu.edu

S. An Blevis et al. (eds.), *Seeing Interaction Design*, Design Research Foundations,
https://doi.org/10.1007/978-3-032-14200-9_9

an Asian student. The foreground signs pictured are by Douglas Hofstadter and are *ambigrams* (Hofstadter, 1987) for Unity and Tolerance, Respect and Peace—appearing here with permission. Ambigrams are calligraphic Artwork which make sense when flipped 180 degrees in orientation.

During the pandemic, anti-Asian racism and hate crimes rose sharply in the USA. It reminds us that the long genealogy of anti-Asian violence in the West that reaches as far back as the nineteenth century is unfortunately not over. Unity, Tolerance, Respect and Peace is the only hope to end anti-Asian racism or any other forms of racism and discrimination.

Image 9.2: iPhone Label, Eli Blevis, Photograph, 2017, Image 9.3: mePhone Photographic Design Fiction, 2017 (Satire), Image 9.4: wePhone Photographic Design Fiction, 2017 (satire).

The text on the back of the iPhone 7 reads *Designed by Apple in California. Assembled in China.* It does not just state *Made in China*, like other products as required by law. How may we interpret *Designed by Apple in California, Assembled in China?* We may be intended to believe that the design is better because it comes from California, and that the manufacture is better because it is done in China. We may be intended to believe that even though it is manufactured in China, it is better because it is designed in California. The latter interpretation is troubling because it signals an unequal relationship (disrespect for one party) rather than a partnership.

(Blevis, 2018) argues that *"the use of text in design is a touchstone between design and politics that should be considered thoughtfully as a matter of good design"* and that "*good design entails respect for others.*" As an exercise in *critical design*, the *mePhone* (Image 9.3) is a constructed photographic fiction that illustrates what sort of texts could be even more offensive, as a matter of satire. By contrast, the *wePhone* (Image 9.4) is a constructed photographic fiction that calls for *design for respect.*

Image 9.5 Qualities of Focus: Flowers, Eli Blevis, Photograph, 2017, Image 9.6: Qualities of Focus: Woman at Desk, Shunying An Blevis and Eli Blevis, Photograph, 2017, Image 9.7: Qualities of Focus: Woman with Tablet Computer, Shunying An Blevis and Eli Blevis, Photograph, 2017, Image 9.8: Qualities of Focus: Flower in Focus, Eli Blevis, Photograph, 2017

The first three images in this collection are deliberately and extremely out of focus. These are in fact photographs of an orchid flower, a woman at a desk, and a woman with tablet computer. Image 9.8 is a photograph of orchid flowers in focus—a focused version of the defocused Image 9.5.

The collection invites distinctions between abstract, lossy, real, and artificial. The images are abstract and lossy—the intent is to make them more universal by reducing the specific information content. The images are truthful—they are minimally processed out of camera and labeled truthfully. Here, truth is distinct from detail. If the images had been mislabeled or processed more in post-production tools such as Photoshop, the intent might have been to *create* artificial work or to *create* a falsehood.

Image 9.9 Three Bridesmaids, Eli Blevis, Photograph (Film), circa 1998

This image was taken in 1997 on Devon Avenue in Chicago and was scanned digitally from film negatives hand processed. Three young bridesmaids and their handsomely dressed chaperones circumnavigate the intruding, protruding, and menacing nose of an SUV. This photograph and the three photographs below were published in Blevis, 2014 to picture the theme of *stillness and motion.*

Image 9.10 Parade I, Eli Blevis, Photograph, 2014, Image 9.11: Parade II, Eli Blevis, Photograph, 2014

The two photographs show a group of very similarly dressed men with similar shopping bags parading at some speed through the corridor of an enclosed walkway connecting shopping malls in Hong Kong. The group stands out in the shopping mall as a crowd similarly dressed. But not a single member stands out as they are all similarly dressed. The background of these photographs is perfectly sharp and still, highlighting the movement of the people the more so.

Image 9.12 Apple Store, Hong Kong, Eli Blevis, Photograph, 2014

Pictured is the Apple store in the IFC mall, Hong Kong on the day that a new product had been released. The crowd may be described as a throng—stillness enforced by its volume and yet with Escheresque movement along its patented glass staircase. The vehicles—below the store—are moving, while many of the people are still.

Herd mentality is often used to explain consumer behaviors. Many of us want to be trendy, not only in fashion trends, but also in new technology adoption (cf., Gladwell, 2006; Heshan, 2009). Some technology companies have successfully created *cult* brands.

Image 9.13 View from Hung Hom, Hong Kong, Eli Blevis, Photograph, 2012

This image is constructed with two photographs that were taken on different days in Hung Hom, Hong Kong in 2012: the left on a good air day, the right on a poor air day. A decade later, evidence of rapid climate change becomes even more compelling. Our collective overconsumption in the past century leads us close to the point of peril.

The foreground signs pictured are by Douglas Hofstadter and are ambigrams (Hofstadter, 1987) for Unity and Tolerance, Respect and Peace—appearing here with permission.

9.2 Summary

This gallery is curated from images appearing in more than a decade of engagement with photography as a mechanism of expression in HCI and interaction design. Herein, we privilege notions of design for respect, truth in image making, and a very abstract thesis concerning stillness and motion in photography. The last image in particular—Image 9.13, in its geometric, symmetrical quadrants of urbanity and environmental decay—concludes the collection in the way that the presentation associated with (Blevis, 2014) concludes. That is, *all motion ends eventually in stillness.*

Image 9.1 Signs at a Rally in Support of Asians and Asian Americans and Pacific Islanders Communities, Shunying An Blevis and Eli Blevis, Photograph, 2023

Image 9.2 iPhone Label, Eli Blevis, Photograph, 2017

Image 9.3 mePhone Photographic Design Fiction, 2017 (Satire)

Image 9.4 wePhone Photographic Design Fiction, 2017 (satire)

Image 9.5 Qualities of Focus: Flowers, Eli Blevis, Photograph, 2017

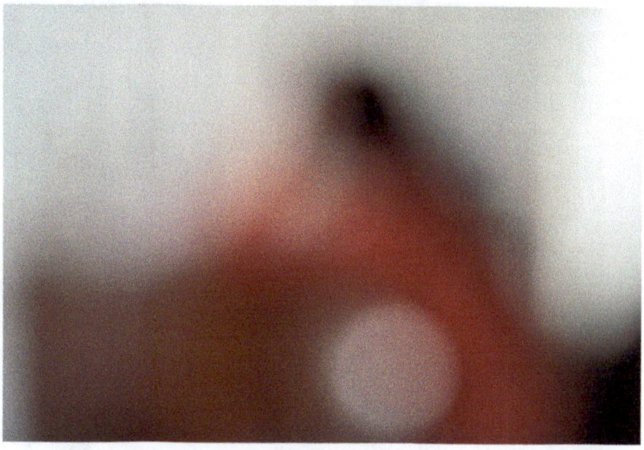

Image 9.6 Qualities of Focus: Woman at Desk, Shunying An Blevis and Eli Blevis, Photograph, 2017

Image 9.7 Qualities of Focus: Woman with Tablet Computer, Shunying An Blevis and Eli Blevis, Photograph, 2017

Image 9.8 Qualities of Focus: Flower in Focus, Eli Blevis, Photograph, 2017

Image 9.9 Three Bridesmaids, Eli Blevis, Photograph (Film), circa 1998

Image 9.10 Parade I, Eli Blevis, Photograph, 2014

Image 9.11 Parade II, Eli Blevis, Photograph, 2014

Image 9.12 Apple Store, Hong Kong, Eli Blevis, Photograph, 2014

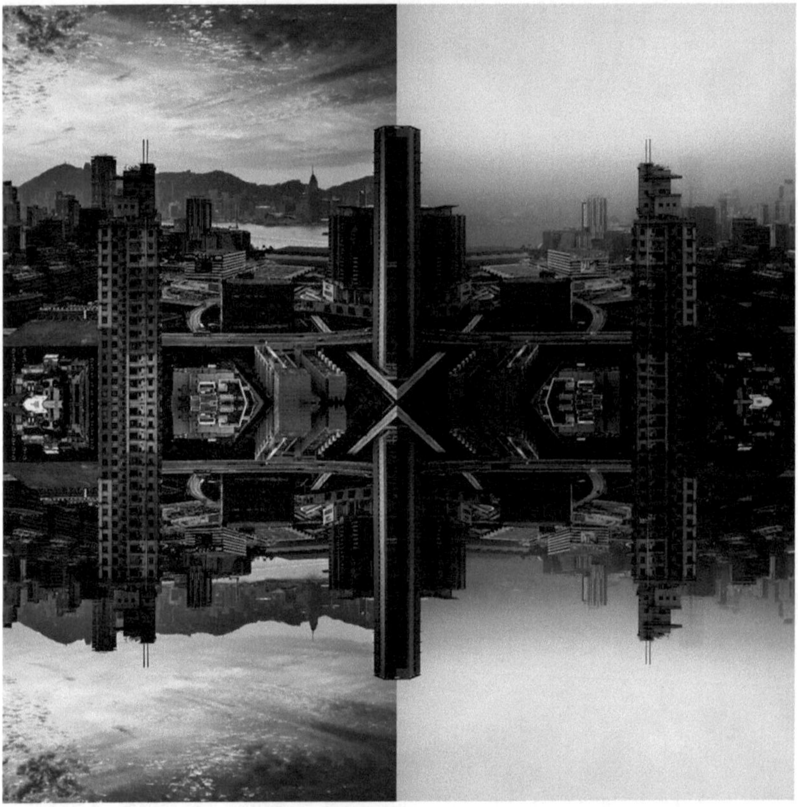

Image 9.13 View from Hung Hom, Hong Kong, Eli Blevis, Photograph, 2012

Acknowledgements Many thanks to the reviewers of the original pictorials and papers upon which this chapter is based.

References

Blevis, E. (2014). Stillness and Motion, Meaning and Form. In *Proceedings of the 2014 Conference on Designing Interactive Systems (DIS '14)* (pp. 493–502). ACM, New York, NY, USA.

Blevis, E. (2017). Qualities of Focus. In *Proceedings of the 2017 ACM SIGCHI Conference on Creativity and Cognition (C&C '17)* (pp. 309–322). Association for Computing Machinery, New York, NY, USA.

Blevis, E. (2018). Seeing What Is and What Can Be: On Sustainability, Respect for Work, and Design for Respect. In *Proceedings of the SIGCHI Conference on Human Factors in Computing Systems (CHI '18)*. ACM, New York, NY, USA.

Blevis, S. A., & Blevis, E. (2022, January–February). Nonlinear Design Thinking: The Limitations of Awareness in the Search for Climate Care. *Interactions, 29*(1), 84–88. https://doi.org/10.1145/3505632

Blevis, S. A., Heidaripour, M., Bardzell, S., & Blevis, E. (2023). Advancing Non-Linear Design Thinking. *Ninth Computing within Limits. LIMITS*. https://doi.org/10.21428/bf6fb269.0b621289

Douglas, R. H. (1987). *Ambigrammi: Un microcosmo ideale per lo studio della creatività*. Hopefulmonster.

Malcolm, G. (2006). *The Tipping Point: How Little Things Can Make a Big Difference*. Little, Brown.

Sun, H. (2009). Understanding Herd Behavior in Technology Adoption and Continued Use: A Longitudinal Perspective. In *DIGIT 2009 Proceedings* (p. 11).

Chapter 10
Fabric, Beads, and Yarn: Crafting (for) Women's Orgasm

Dianya Mia Hua

Abstract This work recognises women's masturbation as a remarkable source of sexual pleasure, a legitimate form of sexual expression, and a means to reclaim women's sexual pleasure and sexuality. This chapter presents five artefacts crafted by women—four of whom are identified as women of colour—in my bedroom for their masturbation. The images are annotated by women's conversations during the craft process to foreground their stories, experiences, and agency embedded in the artefacts.

10.1 Introduction

In the late 1960s women's movement, feminist activists radically challenged the discourse of women's sexuality: "*Sexuality was not just a matter biology, they argue; rather, it was a set of practices and beliefs firmly embedded in and shaped by a complex web of social arrangements, scientific discourses, and gendered power that support the patriarchal status quo.*" (Comella, 2017: 19) In particular, feminist activists re-framed masturbation as a political act for women to take control of their orgasm, reclaim their sexuality, and declare their sexual independence (Comella, 2017; Dodson, 1974; Lieberman, 2017). It was in this context that vibrators were transformed into a political artefact for women's masturbation (Comella, 2017; Lieberman, 2017; Loe, 1998). Although women's masturbation has been entangled with digital technology for more than half a century, works that explicitly address this conjunction remain overwhelmingly marginalised in sexuality research of interaction design and HCI (Bardzell & Bardzell, 2011; Brewer et al., 2006; Carpenter et al., 2018; Kannabiran et al., 2011; Keilty, 2012; Rothstein et al., 2021; Schulte et al., 2021; van Greevenbroek et al., 2023; Zytko & Furlo, 2023).

D. M. Hua
College of Information Science and Technology, The Pennsylvania State University, University Park, PA, USA

D. M. Hua (✉)
School of Design, The Hong Kong Polytechnic University, Tai Kok Tsui, Hong Kong
e-mail: mia.dy.hua@polyu.edu.hk

S. An Blevis et al. (eds.), *Seeing Interaction Design*, Design Research Foundations,
https://doi.org/10.1007/978-3-032-14200-9_10

Building on the feminist framing of women's orgasm, masturbation, and vibrators, this work explores how interaction design can support women's masturbation as a site of activism to shift the politics and power dynamics inside of women's sexuality. This work consists of two sessions. In the first section, we sat in a circle speaking in turn to address our experiences, stories, and perceptions on masturbation. In the second session, we were sitting on the floor, talking, and crafting. Each of us was asked to craft a vibrator individually for our masturbation using materials including fabric, thread, yarns, beads, and cup sequins.

This chapter presents five artefacts crafted by us. Four of us are identified as women of colour, who were not originally from the Northern American or European contexts and have historically been silent and marginalised in sexuality research of interaction design. Hinging upon the notion *the personal is political*, I want to highlight our stories, experiences, and agency embedded in the artefacts. Therefore, I take a slightly uncommon approach to presenting the images: instead of my interpretations, the images are mainly annotated by our conversations during the craft process. Every woman, including myself, will be mentioned under a pseudonym chosen by us.

10.2 Contributions

Drawing on artefacts crafted by women for their masturbation, I hope to amplify the voices of a group of women who have been historically silent and marginalized in sexuality studies in Interaction Design. My second hope is to illustrate women's aspirations of artefacts for their masturbation for future design. I also hope this work could reduce the social stigmas that surround women's masturbation and extend the current scope of sexuality research in interaction design.

10.3 The Heart by Maggie

See Image 10.1.

Last Christmas, a friend of Maggie—who is American and identified as white—gave her a vibrating anal plug as a present. Maggie had been using it as a clitoral vibrator instead. *"I actually have used [the vibrating anal plug] just as a clit vibrator."* said Maggie, *"I was like: they don't even need to designate exactly where this is supposed to be used."* Because that vibrating plug worked very well for her, she researched the company that made it and then purchased a second vibrator. It was a clitoral suction vibrator with a hollow structure which stimulates the clitoris without touching it. *"That thing has changed the game!"* Maggie said enthusiastically, *"Highly recommend!"*

Her vibrator—*The Heart*—was a multiple-purpose vibrator. As the name indicated, it is shaped like a pillowy heart. *"I was thinking that the heart shape might be*

Image 10.1 *The Heart* by Maggie

good because you could use it in like a variety of different ways." said Maggie, *"You could use just the end for more like pointed kind of sensation and you could use it on the vulva or the clit or the butt. But then you might also prefer more surface area with the curve. Also, the heart shape would feel like kind of comfortable to hold."*

The heart came with a wrist strap. Maggie said: *"I was thinking about how you could have a little wrist strap to help you hold it or carry it around."* The idea of carrying the vibrator around with the wrist might be her way to break down the stigmas around women's masturbation and vibrators.

After Maggie shared her craft, Rachel already imagined putting it on a bed. *"If you put it on your bed,"* Rachel said. *"it's like, just so cute!"*

Image 10.2: The Textures of *The Heart.*

Maggie included a clitoral suction structure (the upper left image) in her craft. "*And then ... this kind of reminded me of like the clit sucker."* said Maggie, *"I was imagining you could have even like a clit sucker kind of apparatus."*

Besides, *The Heart* is furnished with different textures for different sensations. *"Some areas are with different textures so that you would like to move this all over the body."* said Maggie, *"It could be just encompassing a ton of different uses."*

Image 10.2 The textures of *The Heart*

10.4 The Dumpling by Michaela

See Image 10.3.

A few months ago, Michaela—who is originally from India and identified was woman of colourr—was subjected to non-consensual sex. Masturbating with a vibrator is her way to reconnect with her body. She shared her experience when we were talking about tattoos.

"I got this [tattoo] recently." said Michaela, *"It's still healing."*

"Was it painful?" asked Anna.

"Um, I shouldn't say what I'm going to say," said Michaela hesitantly, *"I'm not going to answer your question."*

"No, no," Anna assured her.

"I was out all that situation, and I wanted to finally get a tattoo." said Michaela, *"But I had just had my advising meeting and it was the first time I had met a man [after the non-consensual sex] ... On that day, so I was like kind of slightly triggered and dissociated in a closed space. ... So, I was dissociative. But I didn't feel the pain. It felt like: Oh my god, I'm finally back in my body."*

"I know that it sounds like self-harm, but I'm not talking about self-harm." added Michaela, *"It was just like that sensation that woke me up. So, my therapist [told*

Image 10.3 *The Dumpling* by Michaela

me that] the alternative to that is like holding ice in your hand because it gives that sharp sensation to get you back to your body."

"But then now my vibrator helped me to connect with my body," continued Michaela. *"I was like: Good, now I don't need to put some ice in my fridge to just like wake up. Even if I don't orgasm, I use the vibrator to just like come back to my body."*

Michaela's vibrator—*The Dumpling*—is a whole-body vibrator that can be used on the genitals. It is soft and cushiony overall. *"If it's a vibrator,"* said Michaela, *"then you can feel different textures on your body and you might be able to use some of them on your genitals too."*

Image 10.4: Textures of *The Dumpling.*

Dumpling comes with four textures. While two of them offer the sensation of the beard, one simulates the sensation of lips. In doing so, her vibrator enables her to enjoy sensations that a human body might offer without actually interacting with a human.

"This one [the upper left image] is like a beard." said Michaela, *"I really get turned on by the sensation of the beard. I like the beard when a guy kisses me on my body."*

Image 10.4 Textures of *The Dumpling*

"This one [the upper right image] is softer. It feels almost like lips in a way" said Michaela, *"This is a little* harsher *[the lower left image], like a trough stubble. And those [the lower right image] are just regular vibratory beads."*

Maggie was amazed by Michaela including the sensation of beard in her craft. *"It is very interesting that I have never heard people talking about that in terms of sensations of someone's hair crossing me."* said Maggie, *"But it is such a part of [the sexual experience]".*

After Michaele finished her shearing, Maggie admired the sensation of beard in Michaele's artefact again.

"It's so interesting!" Maggie exclaimed, *"I've never really heard people talk about that in terms of the sensation of somebody's hair. But it's like totally such a big part of it."*

"Yeah, immersive experience." responded Michaele, *"This one [the lower left image] ... feels more like stubble, like fresh shave for a day. And this one [the upper right image] is like a beard that's grown for some time. ... I love the sensation of beard on my face and my neck."*

"This is amazing," said Maggie again.

10.5 The Essential by Rachel

See Image 10.5.

Rachel's—who is originally from Indonesia and identified as woman of colour—first encounter with vibrators is like an adventure. *"[I have] only one [vibrator]."* said Rachel, *"I need to go abroad for that. Because in my country, you just have to keep quiet about that. So, I didn't have the courage to buy online or something. When it comes to our house, everyone will be like: what is that? There's a chance for me that year to go to another country. ... That's the time when I try to figure out which one."*

From that day onward, masturbation with vibrators gave her a sense of victory. *"It feels that after twenty or even after thirty we have the chance for that."* said Rachel, *"So, before that, ... we only used our fingers. But then, the feeling is just victory! It's very beautiful how masturbation feels."*

Her vibrator came with two different vibrating units and a button.

"This is the button," said Rachel pointing at the stature in the lower images.

"This is a stimulator that mimics the tongue. If you the oral sensation [you can use it]." said Rachel, *"This one is softer. It is up to you on which one you want to use for your front side or your middle side."*

Image 10.5 *The Essential* by Rachel

By the front side, Rachel means the clitoris. The middle side, the vaginal opening area.

"It also feels like a panty liner, then you can also wear it." commented Anna, *"I feel it's very wearable. Like you could just put it in the inside of your underwear."*

"Yeah, it's also very compact." agreed Rachel, *"You can bring it. It will be nice to put in a small makeup bag, or a small punch."*

"I think you can call it The Essential." suggested Anna.

10.6 Three in One by Anna

See Image 10.6.

Anna believed that designing and marketing for women creates a safe space for women to access sexual pleasure.

"It's a good thing for vibrators to be for women," said Anna, *"because it's a little harder for women to access sex in comparison to men in lots of cultures."*

Since she is from Kenya, Anna specifically addressed women's experiences in African countries.

"In African countries, it's not deemed as a space for women to lead or to ask." continued Anna, *"... I feel like sex is such an invasive thing sometimes. When you think about it, you have to let someone into your body. It's very different for you. It might be easier to do it with a toy than someone because it might be not what is right for you to experience."*

Her artefact is the only penetrative vibrator. As the name indicated, Anna's vibrator—*Three-in-One*—is a tri-vibrator, which stimulated the clitoris, vagina, and anus simultaneously.

Image 10.7: The Details of *Three-in-One.*

In first section of this work, we had a brief discussion about that dildoes are oftentimes oversized. Anna brought it again in the craft workshop.

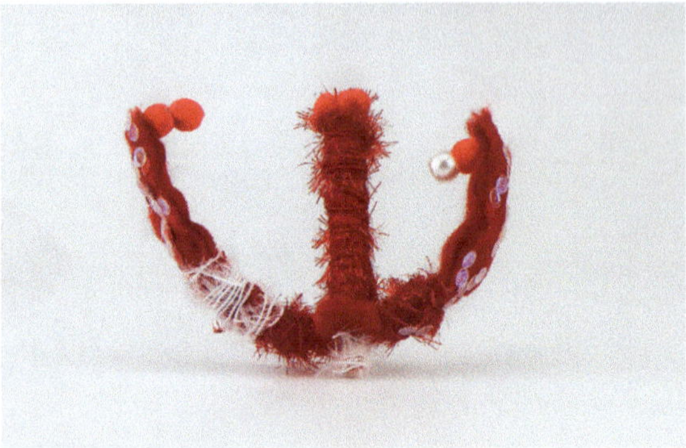

Image 10.6 *Three in One* by Anna

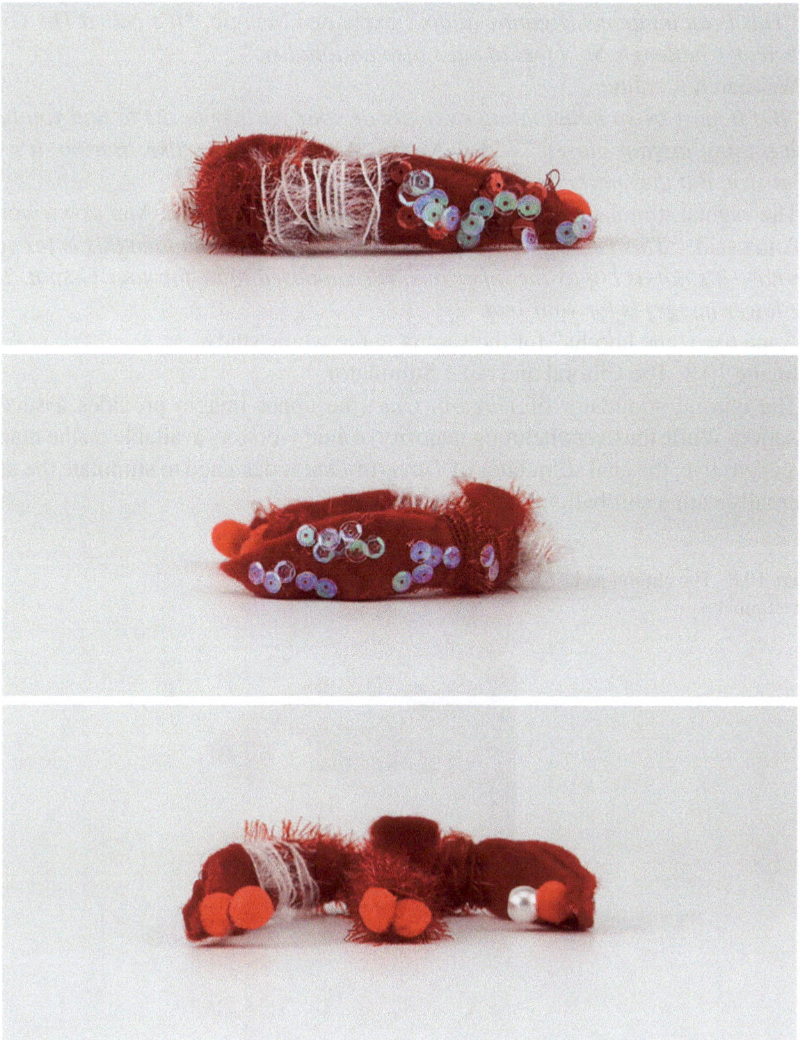

Image 10.7 The details of *Three-in-One*

"Remember yesterday you were saying how dildos were usually so big." said Anna, *"Why did people start doing that?"*

"I don't know." answered Deryn, *"I feel like it's related to men's pride, and the big dildoes were more like men's imagine of what women want."*

"You know about The Great American Challenge? Do you know about that?" asked Maggie.

"No." answered others.

"This is an infamous, gigantic dildo." explained Maggie, *"It's called The Great American Challenge. So, I feel like it's also nationalist."*

We search it online.

"*But it must be so intimidating as a guy or your partner to see to find you have such a thing in your closet."* added Anna, *"And then you're like: But no, it's not about size. But they never believe you."*

The vaginal stimulator in the *Three-in-One* was *"not so big"* in Ann's own words.

Anna said: *"This [the upper image] is like for your clit and then this is for your hoo-ha – it's not as big as the other one. It's supposed to be for your G-spot. This [the lower image] is for your butt."*

Anna used "the hoo-ha" for the vagina in the whole study.

Image 10.8: The Clitoral and Anal Stimulator.

The clitoral stimulator of *Three-in-One* (the upper image) provides a suction sensation. While the overwhelming majority of anal vibrators available on the market are penetrative, the anal stimulator of *Three-in-One* is designed to stimulate the anus externally with a soft ball.

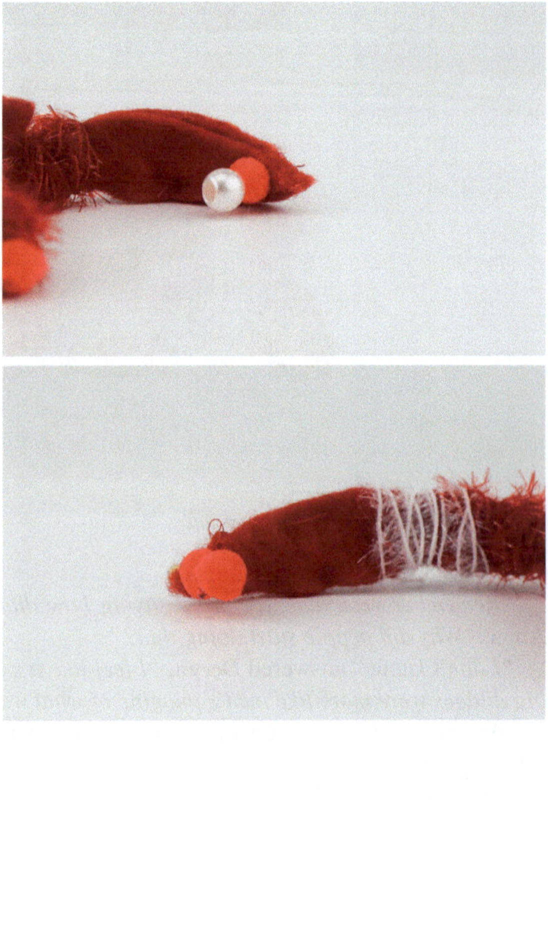

Image 10.8 The clitoral and anal stimulator

10.7 Softy by Deryn

See Image 10.9.

Deryn has an uncommon technique for masturbation. Although she had tried many vibrators, Deryn had not found a vibrator that could fit her masturbation practices.

"I never touch myself." explained Deryn, *"I will lay down and cross my legs. And then I will just squeeze my legs, so it causes pressure on my clitoris."*

Deryn's masturbation practice inspired her to craft an underwear-shaped vibrator named *Softy*. After Deryn told the others that she wanted to make underwear, we discussed the current concept of vibrating underwear.

"I think I am going to make an underwear," said Deryn.

"There is actually an underwear with a vibrator." said Racheal.

"Yeah, I know those vibrating panties, but I hate the concept in general." said Deryn, *"Those vibrating panties always come with a remote control. They are usually designed for the public play and the control is usually in the man's hands."*

"Yeah, I didn't like the cover of the box when I saw it." said Anna, *"[It shows] a man sitting on a couch with the remote control."*

"Ugh, it's not good," said Michaela.

"We have to own the remote!" said Maggie.

Image 10.9 *Softy* by Deryn

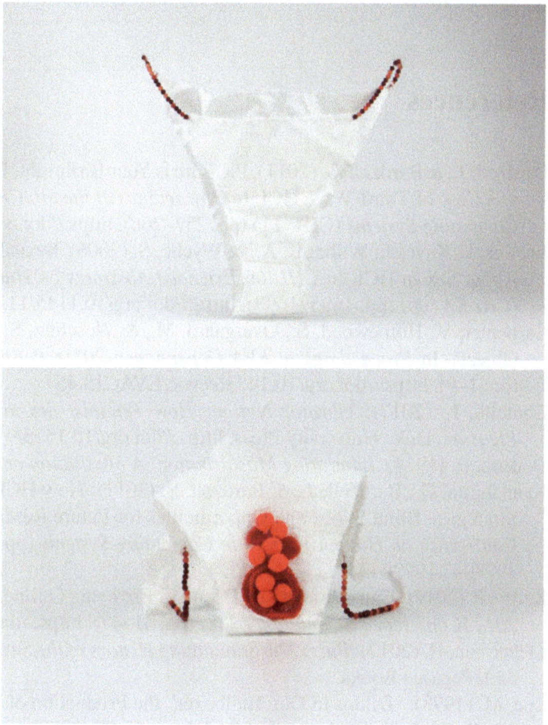

Different from the available vibrating panties, *Softy* does not come with a remote control, and it is only for women's masturbation.

"*I hope it is pretty so that you can feel like I'm sexy when you wear it.*" said Deryn, "*Its outside is very simple but there is a soft structure inside. Even when you cross your leg tightly, you won't feel uncomfortable.*"

"It looks like eggs in a nest," said Maggie.

10.8 Summary

This chapter presents artefacts crafted by women—including myself—in my bedroom for their masturbation. The images are annotated by their conversations during the craft process to highlight women's stories, lived experiences, and beliefs embedded in the artefacts. The intent is for the viewer/reader to interpret each artefact after reading each woman's stories in their way.

Acknowledgements I gratefully thank the contributing editors who have encouraged me to compose this chapter, namely Shunying An Blevis, Miriam Sturdee, and Eli Blevis. I also thank my advisor Shaowen Bardzell as well as other contributors for their valuable comments and feedback. I especially thank the women who participated in my study.

References

Bardzell, J., & Bardzell, S. (2011). Pleasure is Your Birthright: Digitally Enabled Designer Sex Toys as a Case of Third-Wave HCI. In *Proceedings of the SIGCHI Conference on Human Factors in Computing Systems* (CHI '11) (pp. 257–266). https://doi.org/10.1145/1978942.1978979

Brewer, J., Kaye, J., Williams, A., & Wyche, S. (2006). Sexual Interactions: Why We Should Talk About Sex in HCI. In *CHI '06 Extended Abstracts on Human Factors in Computing Systems (CHI EA '06)* (pp. 1695–1698). https://doi.org/10.1145/1125451.1125765

Carpenter, V., Homewood, S., Overgaard, M., & Wuschitz, S. (2018). From Sex Toys to Pleasure Objects. In *Proceedings of EVA Copenhagen 2018: Politics of the Machines—Art and After* (pp. 1–9). https://doi.org/10.14236/ewic/EVAC18.45

Comella, L. (2017). *Vibrator Nation: How Feminist Sex-toy Stores Changed the Business of Pleasure*. Duke University Press. https://doi.org/10.1515/9780822372677

Dodson, B. (1974). *Liberating Masturbation: A Meditation on Self Love*. Bodysex Designs.

Kannabiran, G., Bardzell, J., & Bardzell, S. (2011). How HCI Talks About Sexuality: Discursive Strategies, Blind Spots, and Opportunities for Future Research. In *Proceedings of the SIGCHI Conference on Human Factors in Computing Systems* (pp. 695–704). https://doi.org/10.1145/1978942.1979043

Keilty, P. (2012). Embodiment and Desire in Browsing Online Pornography. In *Proceedings of the 2012 iConference (iConference '12)* (pp. 41–47). https://doi.org/10.1145/2132176.2132182

Lieberman, H. (2017). *Buzz: The Stimulating History of the Sex Toy* (First Pegasus Books hardcover ed.). Pegasus Books.

Loe, M. (1998). "Dildos in Our Toolboxes" the Production of Sexuality at a Pro-sex Feminist Sex Toy Store. *Berkeley Journal of Sociology, 43*, 97–136.

Rothstein, N. J., Connolly, D. H., de Visser, E. J., & Phillips, E. (2021). Perceptions of Infidelity with Sex Robots. In *Proceedings of the 2021 ACM/IEEE International Conference on Human-Robot Interaction (HRI '21)* (pp. 129–139). https://doi.org/10.1145/3434073.3444653

Schulte, B., Søndergaard, M. L. J., Brankaert, R., & Morrissey, K. (2021). Utopian Futures for Sexuality, Aging, and Design. *Interactions, 28*(3), 6–8. https://doi.org/10.1145/3460204

van Greevenbroek, R., Patel, D., & Singh, A. (2023). "Like a Candy Shop with Forbidden Fruits": Exploring Sexual Desire of Cohabiting Millennial Couples with Technology. In *Proceedings of the 2023 ACM Designing Interactive Systems Conference, 1842–1860.*

Zytko, D., & Furlo, N. (2023). Online Dating as Context to Design Sexual Consent Technology with Women and LGBTQ+ Stakeholders. In *Proceedings of the 2023 CHI Conference on Human Factors in Computing Systems* (pp. 1–17).

Chapter 11
Fluid Speculations

Paulina Yurman

Abstract This chapter presents my experimentations drawing artefacts and technologies. The use of watercolour, a fluid medium of leaky boundaries, presented opportunities for defamiliarization, ambiguity and narrative. The inexact quality of the material challenged the rational precision that many technological artefacts are meant to embody and pushed boundaries in conventional forms of artefact representation. Suggesting speculative concepts, these drawing experiments have both utilized and subverted creative processes involved when designing artefacts, inviting thoughts about the experiences the drawn objects evoke. These visual examples present how both drawing and the materiality of the drawing have enabled a space for thinking about artefacts, with the left traces of the actions that occurred in such practices.

11.1 Introduction

The spectrum of activities that encompass design is diverse, yet there seems to be a general consensus that the visualization of ideas is essential. Interest in drawing as a form of discovery and knowledge production is an established body of work in the investigation of creativity, cognition and design processes. Drawing has been described as a visual exploration of ideas that is both creative and communicative (Garner, 1992; Mäkelä et al., 2014; Tversky, 1999), a conversation between the drawing and the individual (Schön, 1983) or as a space where thinking is presenced (Rosenberg, 2008), and is also valued in a wide range of disciplines such as medicine, mathematics, biology and physics (Kaiser, 2005; Kearns, 2019; Penrose, 2015; Wittmann, 2013). Drawing remains an essential practice for many designers, artists, crafts people, architects and engineers, facilitating creativity, communication and processing of information. Different materials and styles used for visualising concepts will have an impact on the families of ideas they can suggest. This chapter presents alternative ways of representing artefacts and design concepts through the

P. Yurman (✉)
University of the Arts London, London, UK
e-mail: p.yurman@csm.arts.ac.uk

S. An Blevis et al. (eds.), *Seeing Interaction Design*, Design Research Foundations,
https://doi.org/10.1007/978-3-032-14200-9_11

use of the fluid medium of watercolour, that enabled defamiliarization and ambiguity and visually suggested possible narratives of use. Drawing enabled a space for thinking about designed artefacts and their associated experiences, where watercolour left a trace of the actions that occurred in such practice and ambiguity became a useful resource.

11.2 Ways of Seeing

In the design of artefacts, drawing can enable manipulation through representation rather than through actual, physical artefacts (Goel, 1995), facilitating a play of ideas before committing to a specific direction. Open ended and suggestive drawings can be particularly important in the early stages of a design project: vagueness can encourage free associations (Tversky & Suwa, 2009). In the design and research of artefacts, ambiguity can promote innovation through the fresh reinterpretation of what might have become familiar. Defamiliarization in design research (Bell et al., 2005) has offered useful examinations of the ways in which we narrate domestic technologies, proposing literary techniques that interrogate perceptions of the familiar. Here, I have used drawing—another form of telling, as Ingold (2013) suggests—for defamiliarization, with watercolour's liquid materiality as a core participant in my approach. Watercolour's fluid and imprecise quality has enabled me to challenge dominant forms of drawing and conceptualising artefacts, and its ambiguous characteristic has been particularly useful for inviting free associations and interpretations.

As someone with experience in the design and realisation of artefacts, I acknowledge the importance that precision has in design processes. In a drawing, precision can be critical for minimizing weight or waste in manufacturing or prototyping, and standardization in drawing is essential when communicating requirements to third parties accurately. However, drawing standards can also prioritize particular ways of seeing that are not necessarily universal. Whilst the effective communication of details in a design idea relies on agreed forms of visualisation that need to be easily understood by a diverse number of stakeholders, they often also need to be adapted according to the context in which they are presented. For example, in her doctorate research collaborating with artisan women in rural communities in Oaxaca, designer Liliana Ovalle often complements her design drawings with photos of plasticine models, anticipating how they might most easily be interpreted by the artisans, who are used to working materials directly with their hands and without the intermediary of the drawing.[1] Another example is when technical drawings of prototypes are adapted to facilitate their reading by omitting standard diagonal hatch lines usually drawn to indicate cross sections, as the pattern can sometimes be misinterpreted as ornamental, resulting in models produced with diagonal lines etched on their

[1] Anecdote from a conversation with Liliana Ovalle, www.lilianaovalle.com.

surface.[2] This adaptation of drawing is a reflection of the different ways in which possible artefacts can be imagined and how the individuals imagining them might choose to best communicate such visions. In historical patent drawings for example, standards for clarity and ease of reproduction still enabled illustrators some artistic license in visualization styles (Prince, 2011), tightly interrelating the idea and its representation into one.

The drawing of an artefact can allow specific qualities to be discussed, and the materials, tools and styles used to generate the drawing can impact on such discussions but can also influence on how the person who drew is perceived. Barnhart and Walkers (2018) researched styles and perceptions of drawing in female and male industrial design students at Iowa State University and found that the idealization of the 'hot sketch', a style often done with marker pens at angular perspectives and more common practice with male students, was associated with the image of the 'competent' designer. Female students, having predominantly trained in fine art style drawing, were often perceived as less confident or less competent. According to the authors, such differences in perception contribute to gender disparity in industrial design practices.

Challenging prevalent forms of design representation can allow room for fresh perceptions of what might have become familiar or dominant. The ways in which we imagine proposals are influenced by our situated perspectives; our subjectivities and personal histories shape how we view and understand things. In my particular case, a female industrial designer born in South America but trained in design and research in the UK, I carry ways of seeing containing a history of multiple cultures and migrations (both geographical and disciplinary) that affect how I see, draw and imagine objects. Growing up in the 'southern cone', technological artefacts often represented symbols of status to be consumed rather than understood, imported from afar with promised associations of progress and incorporated into local kaleidoscopic realities, narratives and ecologies. Whilst much of the ideation in design work is rooted on northern western notions of quotidian life, a critical lens can enable a view of artefacts as complex entities that raise fears and fantasies fed by particular imaginations, narratives and diverse material cultures.

If, as John Berger observes, '*a drawing of a tree shows, not a tree, but a tree being-looked-at*' (Berger, 2005), then the drawings I present here show how I looked at artefacts through a process that merged knowledge about them as they currently exist with ideations about how they might also be understood (or misunderstood). This blurring of boundaries can reflect ways of thinking about artefacts that are not static but in permanent flow, with watercolour offering opportunities for engaging with such fluidity. Drawing in watercolour facilitated a recognition that artefacts do not exist in isolation in unchanging and impermeable states, they are vulnerable to different ways of narrating, perceiving, using and misusing them. The unpredictability that occurred as watercolour fluidly moved on the paper became a useful parallel of how thoughts and ideas evolve and mutate, often in surprising directions.

[2] Personal anecdote, from my experience working in a design studio in London that collaborated with factories in northwestern India.

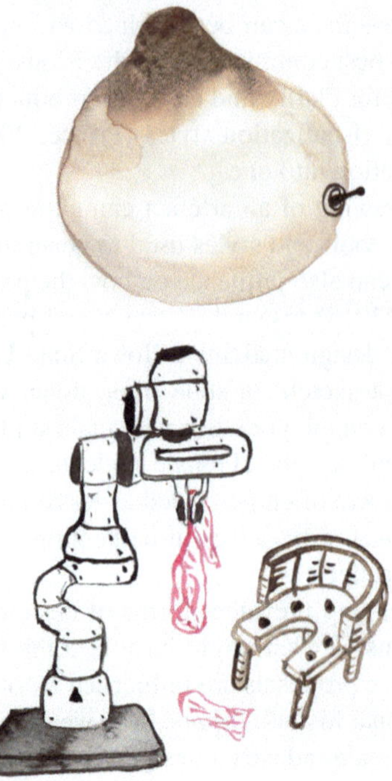

Image 11.1 Drawing artefacts in watercolour invited ambiguity of contexts and use. Paulina Yurman, 2021. Organic machine; Robot and medieval birthing stool. Watercolour on paper

Images 11.1, 11.2, 11.3 and 11.4: Engaging with ambiguity, narrative and free associations.

Watercolour was used not to provide accurate descriptions of artefacts, but rather to open conversations where objects became forms of discourse. The ambiguity of watercolour rendered them as strange yet familiar, often fusing familiar domestic contexts with new technologies. Defamiliarization also occurred when seemingly unrelated objects were drawn merged together, watercolour blurring the boundaries where one ended and another started. Free associations of separate artefacts invited reflections about, for example, certain technologies entering spaces that are sensitive or emotionally charged.

Images 11.5 and 11.6: Adorning the machine.

Watercolour has often been used for the design and illustration of textile pattern and adornment, practices frequently linked to female labour and domesticity. Frills, chintz and ornamentation became associated with stereotypical ideas of 'feminine taste' placed in contrast with the harshness of the industrial world associated with machines, public and male spheres (Parker, 1984; Sparke, 1995). Images 11.5 and 11.6 show how I used watercolour drawing to explore ideas about adorning, feminising, softening or domesticating contemporary machines, fusing the new and unfamiliar with the familiar and nostalgic. Through these drawings I speculate about existing material culture practices of 'dressing' machines such as TV sets, fridges,

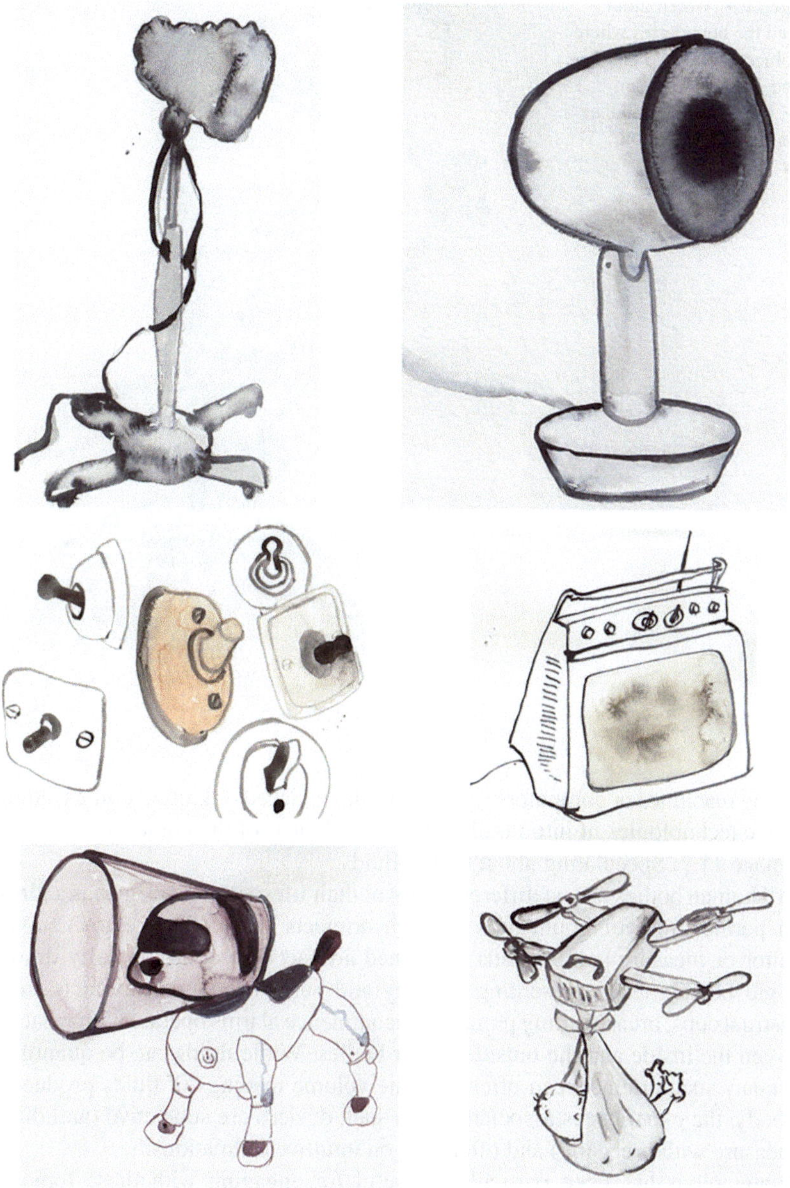

Image 11.2 Drawing artefacts with watercolour invited narrative. Paulina Yurman, 2022. Various technological artefacts. Watercolour on paper

Image 11.3 Watercolour blurred the boundaries where one object ended and another started. Paulina Yurman, 2021. Pram and surveillance camera. Watercolour on paper

washing machines or computers with embroidered, laced or knitted doilies, attempts to make technologies fit into familiar, traditional ideas of the home.

Image 11.7: Speculating about bodily fluids.

All human bodies leak at different times of their lifespan and design has addressed such permeability of boundaries through artefacts that collect, extract, contain, monitor or measure bodily fluids. Designed artefacts are conventionally drawn as of rigid boundaries, representing stability and permanence, yet artefacts such as menstrual cups, breastfeeding pumps or incontinence alarms operate with what flows between the inside and the outside of our bodies. While fluids can be quantifiable, and many such artefacts can offer accurate volume readings of fluids produced by the body, the experiences associated with such devices are subjective (and difficult to measure with precision) and often rely on intuitive estimations.

Watercolour has been particularly useful for engaging with these topics and for challenging the precision that many technological artefacts are meant to embody. The boundary blurring of watercolour and its fluid, unpredictable movements on paper as pigment and water moved and suggested forms impacted on the content of the drawings, an engagement with fluidity itself.

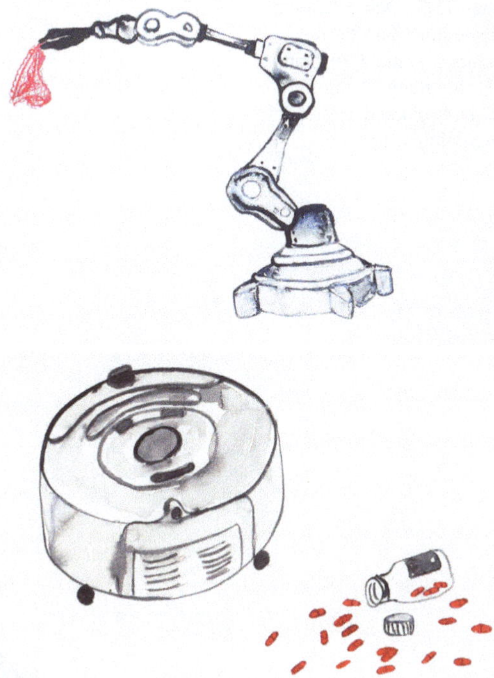

Image 11.4 Various interpretations of the drawings can emerge in the space between what is drawn and what is left out. Paulina Yurman, 2022. Robot with intimate garment; Robot vacuum cleaner and spilled pills Watercolour on paper

11.3 Summary

In this chapter I have shared ideas about fresh ways in which artefacts and technologies can be represented and imagined, two aspects of design research work that I believe are interwoven. I have also described how watercolor impacted on my representation of speculative concepts and narratives associated with the objects drawn. While the text is only a partial account of the thinking in action that took place in the process, the drawings presented here are contributions on their own right. The annotations accompanying the drawings are meant to provide background about the thoughts that emerged while and after they were created, but readers are encouraged to view them through their own interpretations.

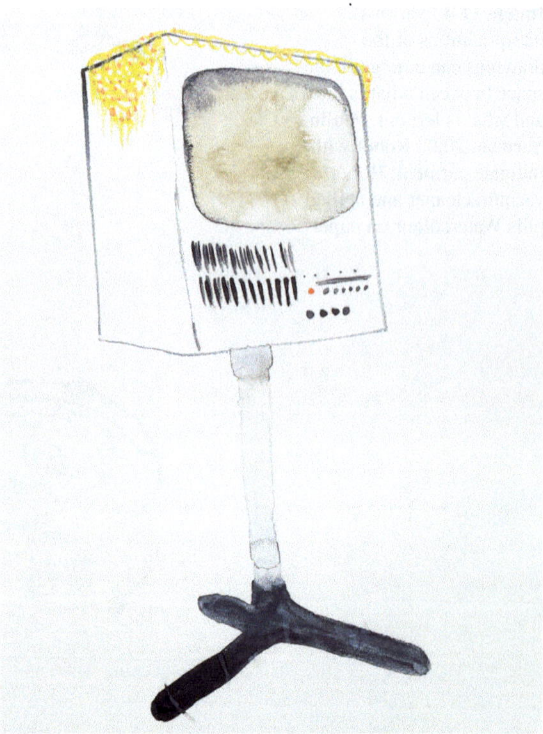

Image 11.5 Domesticating, ornamenting, familiarizing machines. Paulina Yurman, 2021. Standing TV monitor with embroidered doily. Watercolour on paper

Image 11.6 Domesticating, ornamenting, familiarizing machines. Paulina Yurman, 2022. Various technological artefacts adorned with lace and embroidery. Watercolour on paper

Image 11.7 The fluidity of watercolour enabled thoughts about artefacts that deal with bodily fluids. Paulina Yurman, 2021. Artefacts for leaky bodies. Watercolour on paper

References

Barnhart, B., & Walters, K. (2018). Perpetuating the Dominance of the Male Industrial Designer. In *International Conference on Engineering and Product Design Education*, 6 & 7 September 2018, Dyson School of Design Engineering, Imperial College, London, UK.

Bell, G., Blythe, M., & Sengers, P. (2005, June). Making by Making Strange: Defamiliarization and the Design of Domestic Technologies. *ACM Transactions on Computer-Human Interaction (TOCHI),* 12(2), 149–173. https://doi.org/10.1145/1067860.1067862
Berger, J. (2005). *Berger on Drawing.* Occasional Press.
Garner, S. (1992). The Undervalued Role of Drawing in Design. In D. Thistlewood (Ed.), *Drawing Research and Development.* Longman Group, UK Limited.
Goel, V. (1995). *Sketches of Thought.* The MIT Press.
Ingold, T. (2013). *Making: Anthropology, Archaeology, Art and Architecture.* Routledge.
Kaiser, D. (2005). *Drawing Theories Apart: The Dispersion of Feynman Diagrams in Postwar Physics.* University of Chicago Press.
Kearns, C. (2019). Is Drawing a Valuable Skill in Surgical Practice? 100 Surgeons Weigh in. *Journal of Visual Communication in Medicine, 42.*
Mäkelä, M., Nimkulrat, N., & Heikkinen, T. (2014, February). Drawing as a Research Tool: Making and Understanding in Art and Design Practice. In *Studies in Material Thinking* (Vol 10). AUT University
Parker, R. (1984). *The Subversive Stitch. Embroidery and the Making of the Feminine.* I.B. Tauris & Co Ltd.
Penrose, R. (2015). *How Drawing Is Used for Maths and Science.* Video. www.youtube.com/watch?v=hb4V4O-7M4s
Prince, K. (2011). *The Art of the Patent.* Glassko Press.
Rosenberg, T. (2008). New Beginnings and Monstrous Births: Notes Toward an Appreciation of Ideational Drawing. In S. Garner (Ed.), *Writing on Drawing: Essays on Drawing Practice and Research* (pp. 97–110). Intellect Books.
Schön, D. (1983). *The Reflective Practitioner: How Professionals Think in Action.* Basic Books.
Sparke, P. (1995). *As Long as It's Pink. The Sexual Politics of Taste.* HarperCollins.
Tversky, B. (1999). *What Does Drawing Reveal About Thinking?* Stanford University.
Tversky, B., & Suwa, M. (2009). Chapter 4 Thinking with Sketches. In A. Markman, & K. Wood (Eds.), *Tools for Innovation.* https://doi.org/10.1093/acprof:oso/9780195381634.003.0004
Wittmann, B. (2013). Outlining Species: Drawing as a Research Technique in Contemporary Biology. *Science in Context, 26*(2), 363–391. https://doi.org/10.1017/S0269889713000094

Part IV
Inspirations

Chapter 12
Inspirations: Meaning and Form, Visual Thinking Gallery

Shunying An Blevis and Eli Blevis

Abstract This chapter on inspirations shares a curation of images from our earliest work on making photographs be objects and embodiment of design knowledge in and of themselves in interaction design and HCI. The earliest paper appeared in *interactions* in 2011, wherein it introduced the idea of photograph as article to that periodical as a regular contribution from then until now. The photographs were not always obviously related to HCI—an aspect of the work that we and luckily the *interactions* editors at the time—Jon Kolko and Richard Andersen—considered to be a feature.

12.1 Contributions

The selection of photographs here are curated to represent our early work in several contexts, from (a) unintentionally ironic signage, to (b) contrasts in interfaces between digital and predigital things, to (c) gambling as a form of futurism, to (d) the semiotics of unusual dogs, to (e) traditional crafts, street vending and sustainability, to (f) collectivism and capitalism, to (g) the importance of "a more physically connected future."

Image 12.1 Caution, Eli Blevis, Photograph, 2011

At the entrance of a restroom, a vinyl applique restroom sign is missing leg and an arm. Below it, a bright yellow with black text folding sign warns *"Caution, Attention"* to signal a newly washed, still wet floor. The figure on the restroom sign was not meant to appear this way—it has been torn. The juxtaposition of the two signs in the photograph is surprising and humorous to some degree. The wet floor warning sign

Photographs in this chapter are selected from *Digital imagery as meaning and form in HCI and design: an introduction to the Visual Thinking Backpage Gallery* (Blevis, 2011), *The Visual Thinking Gallery: A Five-Year Retrospective*, (Blevis, 2016), as well as some new photographs not previously published.

S. A. Blevis · E. Blevis (✉)
Luddy School of Informatics, Computing, and Engineering, Indiana University, Bloomington, IN, USA
e-mail: eblevis@iu.edu

S. An Blevis et al. (eds.), *Seeing Interaction Design*, Design Research Foundations,
https://doi.org/10.1007/978-3-032-14200-9_12

has inadvertently become more effective because the torn figure reminds people of the potential consequence of not exercising caution on wet floors.

Image 12.2 Interface, Then and Now, Eli Blevis, Photograph, 2012

In the center of this photograph is a round 24 hour rare Universal Genève watch dial from the 1960's or 1970's in the collection of the authors. Evenly spaced markers form a large circle on the dial, indicating time (i.e., 24 hours—marked with numerals and long-ago expired tritium squares, and a shared scale for 60 min, 60 s—marked in tenths). The background of the half of the large circle is in black while the other half in white, indicating day and night. This watch dial design embodies our cognitive understanding of time as being circular (c.f., Fauconnier & Turner, 2008): the watch's second hand in the leftmost sub-dial makes one rotation every 60 s (i.e., one minute), its minute hand makes one rotation every 60 min (i.e., one hour), while its hour hand unconventionally makes one rotation every 24 hours, rather than every 12 hours as is more typical.

The dial also has three sub-dials. One is for the second hand. Two are for a chronograph function. In the bottom most dial is an arrow pointing down to North as a sun compass function. The arrow points down, because on this dial design, 12 is at the top and 24 at the bottom. With a 24-hour watch dial, there are two common designs—12 at the top, or 24 at the top. This 12 at the top design has the cognitive mapping effect of following the movement of the sun, as it rises from night until day and back again to the night.

The dial as pictured rests on top of a plastic holder purpose design for watch repairs. This holder in turn rests on the box from an Apple iPhone. The dial is the *then*. The iPhone is the *now*. For some, the iPhone is an obvious improvement. For others, this old watch dial is a masterpiece of interface design sought after by collectors and people who just like the aesthetics of vintage.

Image 12.3 Guard Dogs, Shunying An Blevis and Eli Blevis, Photograph, 2012, Image 12.4: Lapdog Bag, Eli Blevis, Photograph, 2012

Two women in uniform patrol the mall with two mid-sized security dogs. The dogs look friendly rather than fierce—not security in the sense of guarding, but rather security in the sense of emotional well-being. The two women are also smiling and friendly—also security in the sense of bringing joy to passersby.

In 2012, a man carries a small dog in a bag in Hung Hom, Hong Kong. In 2011, Intel trademarked and released its *Ultrabook*. As a humorous twist, this photograph was titled *Ultrabark Lapdog Bag* when it first appeared in the Visual Thinking Gallery.

Humans have developed complex relationships with animals throughout history. Animals, such as dogs, can help us with work, provide us with companionship, and more. Animal-Human Interaction (AHI) is an important and interesting field of study. Animals can also facilitate human–human interaction. For example, dogs can serve as conversation starters, as seen in these photographs. As dog owners, we have engaged in many chats with people in our neighborhood, in dog parks, and in other public places when we walk our dogs.

Image 12.5 Good Fortune, Shunying An Blevis and Eli Blevis, Photograph, 2012

A man checks horse racing reports published in local newspaper with a magnifying glass outside a restaurant in Sheung Wan, Hong Kong in 2012.

Psychology research (Darke & Freedman, 1977) finds *that "irrational beliefs about luck can serve as a source of positive expectations for the outcome of future events."* This may help to explain why gambling activities such as lottery and betting on horse racing are so popular. Winning or losing in those activities largely depends on odds. Some can pray for good fortune. Some others do not like to gamble with the future. Some pressing contemporary matters (e.g., climate care) are better not left to chance.

Image 12.6 Traditional Chinese Quilt Making—*Tan Mian Hua*, Shunying An Blevis and Eli Blevis, Photograph, 2012, Image 12.7: Quilt Maker's Friends—*Tan Mian Hua*, Shunying An Blevis and Eli Blevis, Photograph, 2012, Image 12.8: Quilt Maker's Friends—*Tan Mian Hua*, Detail, Shunying An Blevis and Eli Blevis, Photograph, 2012

An older man crafts a cotton quilt in a street shop in Rui An, Zhe Jiang province, China. The shop is old, with exposed bricks on the walls and worn-out wood frames and beams. Next to it, a hardware store also shows its age. Despite the weathered appearance, the quilt maker has everything he needs to make quilts for the local community. He even has some stools for his friends or customers to sit on. Two men of his age sit in front of the shop, while another one watches the quilt making process.

Tan Mian Hua, or Chinese quilt making, is a traditional craft that dates back to 1200 CE. Quilts can be made from new or used cotton (e.g., by remanufacturing an old quilt). The quilt maker in the first photograph uses a relatively modern semi-automated machine.

Tan Mian Hua is becoming a lost craft with rapid urbanization. After all, it is more convenient and economical to buy factory-made quilts from stores or online. However, crafts from the past are often more sustainable alternatives, and can serve as inspirations when we aim to minimize our environmental footprint, according to McGovern, 2008.

Image 12.9 Sharing Photographs—*Tan Mian Hua*, Shunqing An, Photograph, 2012

Street photography as an Artform is oftentimes primarily a social skill. Here the photographer is shown sharing the images above with the people who appear in them. The camera is an agent of human–human interactions. The camera in the picture is a Leica rangefinder—a design that was originally intended to be discreet compared to other cameras. Nowadays, smart phone cameras are discreet. The Leica rangefinder looks very much like an "actual" camera, changing the interactions between photographer and people who appear in photographs.

Image 12.10 Street Vending, Shunying An Blevis and Eli Blevis, Photograph, 2012

A woman fries a popular local radish pastry (*Deng Zhan Gao*) at a food cart on a street corner of Rui An, Zhe Jiang in 2012. She wears a colorful apron with a Snoopy character from the Peanuts comic strip.

Street vending is widespread around the world. It is still an integral part of many urban economies even though it may be more regulated in some countries or regions than some others (Recchi, 2021). It provides employment opportunities and delivers needed services in the communities at low cost. In China, street vending was promoted in late 70's and early 80's last century when China went through economic reform and opened up. It helped to create the economic boom during that period (Song, 2020). In order to recover from the Covid pandemic and resulting economic fallout, the Chinese government has again started to promote street vending (Chan & Zhang, 2023).

Image 12.11 Village Activity Center in Rui An, Zhe Jiang, Shunying An Blevis and Eli Blevis, Photograph, 2012, Image 12.12: Village Activity Center in Rui An, Zhe Jiang, Shunying An Blevis and Eli Blevis, Photograph, 2012

People gather in a village activity center to play Mahjong in Rui An, Zhe Jiang province (China). Among them are men and women of different age. Some are playing, some are watching, and some are waiting for more players to join. All the activities seem to be self-organized, and the facility is collectively maintained.

A portrait of Chairman Mao Zedong and two Chinese flags are hanging on the wall in the first two photographs, indicating the presence of village government and communist leadership on formal occasions in this activity center. In Image 12.11, the woman at the right of the image strikes a pose that is similar to the portrait of Mao Zedong, arms folded and smiling. In the background of the third photograph is a raised stage with a pile of benches stored. According to local residents' account, this village activity center hosts performances for holiday celebrations.

Zhe Jiang province is an epicenter of capitalist development in China. It has the strongest private sector in the country, and among them is the company, Alibaba. By way of contrast, this village activity center is a symbol of Chinese collectivism.

Image 12.13 Social Interaction, Shunying An Blevis and Eli Blevis, Annotated Photograph, 2018

A young girl gently places a butterfly on the arm of an elderly woman in the Niagara Falls Parks Butterfly Conservatory in 2018. Her mother is helping by holding her arm. Her younger sister and other people are watching. The sharing of the butterfly in this scene is an important and heartwarming social interaction. It is the highlight of the trip for everyone involved as captured in the photograph.

This photograph is especially thought-provoking after we all went through a long-time isolation because of the global pandemic of 2020–2022. Digital technologies such as video conference software and other collaboration platforms played significant roles in curbing COVID-19 transmission and shielding productivities (Alghamdi & Alghamdi, 2022). However, recent neuro-science research suggests that social interaction through physical touch as pictured in the photograph is essential for mental wellbeing. It algins with calls for "a more physically connected future" (Chokshi & O'Connor, 2021).

The photographs show two alternative annotations of the scene. In one the children in the picture are grayed out. In the second, a spotlight of colour highlights the butterfly and the interaction with it. Here, the children are diffused with a half-tone pattern filter. The photograph is Artwork. Notwithstanding, we have used filters to hide the faces of the children, while still preserving the meaning of the photograph.

Image 12.1 Caution, Eli Blevis, photograph, 2011

12.2 Summary

As we state above under the heading of *Contributions*, the selection of photographs here are curated to represent our early work in several contexts, from (a) unintentionally ironic signage, to (b) contrasts in interfaces between digital and predigital things, to (c) gambling as a form of futurism, to (d) the semiotics of unusual dogs, to (e) traditional crafts, street vending and sustainability, to (f) collectivism and capitalism, to (g) the importance of "a more physically connected future." These are a curation of photographic Artwork which mostly appeared early in our contributions to the idea of being visual and letting visual materials be forms of knowledge in and of themselves in HCI and its confluence with Design. Thus, we count these among the early forms which we ourselves and many other pioneers have contributed leading to the now reasonably common practices of pictorial essays as an archival form within HCI. The acceptance has been slow and gradual, and yet thankfully continues to foster a community of like-minded researchers, designers, and artists.

Image 12.2 Interface, then and now, Eli Blevis, photograph, 2012

Image 12.3 Guard dogs, Shunying An Blevis and Eli Blevis, photograph, 2012

Image 12.4 Lapdog bag, Eli Blevis, photograph, 2012

Image 12.5 Good fortune, Shunying An Blevis and Eli Blevis, photograph, 2012

Image 12.6 Traditional Chinese quilt making—Tan Mian Hua, Shunying An Blevis and Eli Blevis, photograph, 2012

Image 12.7 Quilt maker's friends—Tan Mian Hua, Shunying An Blevis and Eli Blevis, photograph, 2012

Image 12.8 Quilt maker's friends—Tan Mian Hua, Detail, Shunying An Blevis and Eli Blevis, photograph, 2012

Image 12.9 Sharing photographs—Tan Mian Hua, Shunqing An, photograph, 2012

Image 12.10 Street vending, Shunying An Blevis and Eli Blevis, photograph, 2012

Image 12.11 Village activity center in Rui An, Zhe Jiang, Shunying An Blevis and Eli Blevis, photograph, 2012

Image 12.12 Village activity center in Rui An, Zhe Jiang, Shunying An Blevis and Eli Blevis, photograph, 2012

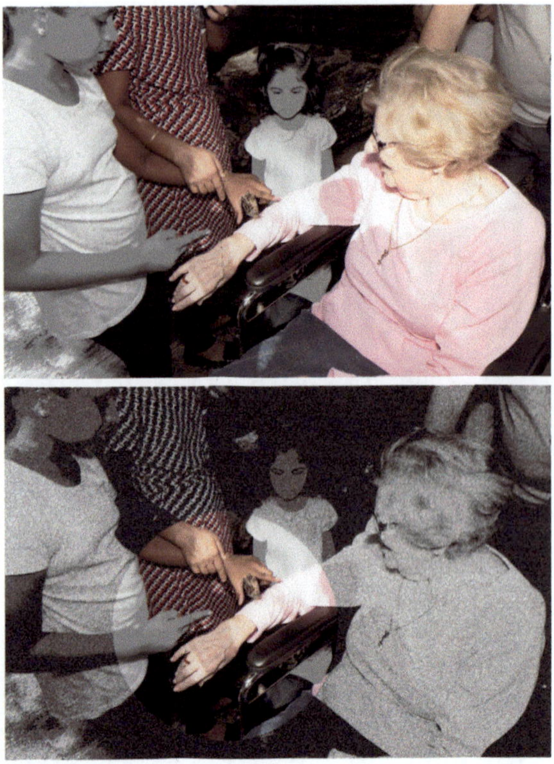

Image 12.13 Social interaction, Shunying An Blevis and Eli Blevis, annotated photograph, 2018

Acknowledgements Many thanks to the reviewers of the original pictorials and papers upon which this chapter is based.

References

Alghamdi, N. S., & Alghamdi, S. M. (2022). The Role of Digital Technology in Curbing COVID-19. *International Journal of Environmental Research and Public Health, 19*(14), 8287.

Blevis, E. (2011, September). Digital Imagery as Meaning and Form in HCI and Design: An Introduction to the Visual Thinking Backpage Gallery. *Interactions, 18*(5), 60–65.

Blevis, E. (2016). The Visual Thinking Gallery: A Five-Year Retrospective. In *Proceedings of the 2016 ACM Conference on Designing Interactive Systems* (DIS '16) (pp. 1096–1110). ACM, New York, NY, USA.

Chan, N., & Zhang, E. (2023, May 29). Hawkers Back on China's Streets as Economic Recovery Teeters. *World at Work*. Reuters.

Chokshi, V., & O'Connor, D. (2021, August). *Social Touch and Our Post-pandemic Future*. www.hopkinsmedicine.org. Retrieved from, December 20, 2023.

Darke, P. R., & Freedman, J. L. (1977). The belief in good luck scale. *Journal of Research in Personality, 31*(4), 486–511.

Fauconnier, G., & Turners, M. (2008). *The Way We Think: Conceptual Blending and the Mind's Hidden Complexities*. Basic books.

Kamal, M. M. (2020). The triple-edged sword of COVID-19: Understanding the Use of Digital Technologies and the Impact of Productive, Disruptive, and Destructive Nature of the Pandemic. *Information Systems Management, 37*(4), 310–317.

McGovern, U. (2008). *Lost Crafts: Rediscovering Traditional Skills*. Chambers.

Recchi, S. (2021). Informal Street Vending: A Comparative Literature Review. *International Journal of Sociology and Social Policy, 41*(7/8), 805–825.

Song, S. (2020). Street Stall Economy in China in the Post-COVID-19 Era: Dilemmas and Regulatory Suggestions. *Research in Globalization, 2*(2020), Article 100030.

Chapter 13
Plurality, Revisited

Miriam Sturdee, Makayla Lewis, Josh Urban Davis, Angelika Strohmayer, Sarah Fdili Alaoui, Nantia Koulidou, and Katta Spiel

Abstract Are we artists or researchers? Makers or practitioners? Do the boundaries shift as we create and grow? This pictorial chapter revisits a collaboration and collective, first bound together by plurality in our shared identities between the academic, educational, industry and self-expression, now reformed and renewed. Some of us pull back from the boundary with research, separating the artistic self from the research 'other'. Some of us find ourselves pulled toward it as it retreats. This collection of images picks up where we left off; the story continues, and we invite you to be part of it. We, seven artists as researchers/artists as researchers, again

M. Sturdee (✉)
University of St. Andrew's, St Andrews, UK
e-mail: ms535@st-andrews.ac.uk

M. Lewis
Kingston University, Kingston Upon Thames, UK
e-mail: m.m.lewis@kingston.ac.uk

J. U. Davis
Dartmouth College, Hanover, USA
e-mail: hello@joshurbandavis.com

A. Strohmayer
Northumbria University, Newcastle Upon Tyne, UK
e-mail: angelika.strohmayer@northumbria.ac.uk

S. F. Alaoui
University of the Arts London, London, UK

Université Paris Saclay, Gif-Sur-Yvette, France

S. F. Alaoui
e-mail: s.fdilialaoui@arts.ac.uk

N. Koulidou
Sheffield Hallam University, Sheffield, UK
e-mail: k.koulidou@shu.ac.uk

K. Spiel
Human-Computer Interaction Group, TU Wien, Vienna, Austria
e-mail: katta.spiel@tuwien.ac.at

S. An Blevis et al. (eds.), *Seeing Interaction Design*, Design Research Foundations,
https://doi.org/10.1007/978-3-032-14200-9_13

bare our souls for you here and curate a selection of our lives and research agendas to be consumed and absorbed.

13.1 Introduction

The first *plurality* was a celebration of our dual natures as artists and researchers, exploring and sharing our experiences, research and artistic outputs, blended in ways which spoke to our methods of working and practising (Sturdee et al., 2021). In the intervening years, our paths have crossed periodically, but our lives and situations have changed. Some of us have moved across—or between—countries to different roles and institutions. What brings us back together is the shared joy of creating and sharing the work we do and the art we make.

This chapter enables us the privilege of speaking to our former selves: who were we then? How have we changed? How has this affected our creative practices? For some, this will have been a logical or calm transition, expanding what we already knew; for others, perhaps fireworks of inspiration, confusion and change. As a reflective practice, the act of putting together this curated snapshot has value in itself—confronting who we have become and what we had hoped to achieve.

We invite you to explore this journey within our practice and research, stopping at each location, interrogating each stylistic choice or nuance, and considering what you value in artistic or research endeavours. May you find meaning, may you find joy, may you find inspiration.

13.2 Contribution

This chapter presents a retrospective of four years of artistic and research practice across seven unique approaches. Each individual continues to work in this space, and their insights and reflections can offer a window into a plurality of practices for those already working in or hoping to engage in similar spaces.

13.3 Carving Out Time

Previously, I focused on older works in the knowledge that I had yet to make time for artistic practice when so much of my creative output focused on furthering my career as an academic. It shamed me to see such a huge part of myself neglected, and despite the continued success of my artistic and sketching self in academia, I hoped that putting this dual nature under the microscope would prompt me to change things. I would like to say I made more time for myself, for art, but I found the boundaries had changed whilst I wasn't looking. New forms of creative expression took over downtime, which allowed me to be away from the desk or computer. I began to bake bread and learn how to cook, pouring attention and time into that instead—because, after all, who doesn't like to eat at the end of the day? However, there was still a niggling doubt that I needed to address this part of myself, long neglected.

The etching press I poured my savings and soul into waits to be loved. When I speak of carving out time, it is in direct reference to this tangible and physical interaction that comes with carving into the lino block or scratching burrs into zinc sheet. At first, it became a Christmas activity, then not even that. The next time I started a lino cut, it was a responsive piece during a workshop on art in HCI (Sturdee et al., 2022), but this also remains unfinished. How might I bring etching into my research? I find myself wondering... Something that might separate the art we do for ourselves from that which we do for research is prompting. The first two images in this section were done for joy –carving out some time for inspiration– if they are now used in a book or paper, they are no less joyful. The tension does not disappear, though; by discussing these works, are they now sullied? Is everything I create now research (Blair et al., 2022)? When do we switch off? My desire to further pull every thread from my practice into research also has its roots in defiance against developments in AI, as I see friends lose money as contracts are not renewed because 'oh, I can just use AI instead'.[1] If I can keep artistic practice, keep it real, and keep it relevant, then nobody can challenge us. Working in a computer science department means that AI is a huge focus in research and teaching; it is the new bandwagon. Close to my heart, though, is the knowledge that some forms of creativity and artistic output can never be digitised or replaced Goldschmidt (2008). Eventually, AI will be another tool in the box. In the meantime, I'll keep trying to carve out some time and lino.

Image 13.1: Old Vines

In 2022, I visited my aunt in France during the winter. She lives near Cognac, and the region is known for growing grapes. Whilst walking out on a dry, overcast day, I was struck by the shapes the vines made. Gnarled and twisting, in infinite detail, like they had been drawn with a fine pen. They reminded me of stricken people, lost in a decaying dance, the organic forms inviting reflection. I took several photographs for reflection, sneaking over the drainage ditches to get closer to the details. I'll probably never get around to drawing these, I thought. I am pleased to be wrong, even in a small part.

Image 13.2: Fika

In the Spring of the following year, I was fortunate to experience living in Sweden. The Swedish people have an active 'fika' culture, where to take a break with coffee and cake is literally 'baked in' to their culture. A colleague and I installed ourselves in one corner of a busy coffee shop for a session on observational drawing, sneaking glances at the people around us. Having always wanted to be the sort of person who sketched in coffee shops, this felt very much like I had unlocked an achievement.

[1] https://amp.theguardian.com/artanddesign/2023/jan/23/its-the-opposite-of-art-why-illustrators-are-furious-about-ai.

Image 13.1 *Old Vines,* Japanese ink on Japanese calligraphy paper (Miriam Sturdee, 2023)

Image 13.2 *Fika*, Copic Marker and Rotring Tikky in Moleskine Sketchbook (Miriam Sturdee, 2023)

Image 13.3 *Dust & Reminders*, Digital photograph, iPhone (Miriam Sturdee, 2023)

Image 13.3: Dust & Reminders

I have long been a fan of etching and linocut, and during lockdown, I decided to realise my dream of owning a press. During the first *Plurality* collaboration, I told myself that I would make more time for my art. However, the press remains the elephant in the room, a reminder that I am not quite there yet. The dust has moved house with me, and an unfinished Christmas card from 2021 peeks out of the basket alongside the tools.

13.4 I Am Taking Back My Sketches

I have spent the last 15 years creating sketch prompts for others to use—participants would colour, rip, and destroy their sketches to craft a visual story reminiscent of their lived experience, e.g. Kimbell et al., 2021. The positive response to Sturdee et al. (2021) from fellow artists and the HCI community reshaped the phrase they used at the start of sketch workshops: "Don't worry, I am not emotionally attached to my sketches. Use them, change them, discard them" (e.g., Kimbell et al., 2021; Lewis & Coles-Kemp, 2014) to "Let's work together, please use the pens to create to share your lived experience about […], I will work on mine alongside you, and then we can share" (e.g., Lewis, 2022b). Their role has adjusted to encourage and inspire others to visualise their stories by sharing theirs. As a result, I now have a collection of sketches representing varied experiences, feelings, and challenges in our ever-changing digital world. From visualising their smartphone use in the workplace and at home, the impact of notifications on health and well-being, and emerging technologies' role in everyday art practice (e.g., Lewis, 2023; Lewis et al., 2022b, 2023a, 2023b; Sturdee et al., 2021) to accept technology as an information and communication lifeline because of their recent epilepsy diagnosis (see Images 13.4, 13.5 and 13.6).

This visual reflection and sharing have sparked a new relationship between art as a researcher/researcher as an artist. My sketches are no longer restricted to the methodology of others' stories but are now a story in their own right. Auto-ethnography (Ellis, 2011) has allowed me to embrace their experiences and feelings through art and accept their creations as valid data that's enough to spark dialogue in the HCI community. This creative freedom has led to the perfection of sketch prompts being 'binned' (gone are the days of their sketches being neat and overly obvious to support participants to craft their stories better, e.g., Sturdee., 2021). Now, I embrace loose mark-making, feeling the page, throwing oneself onto it, being in the now, and not caring about the final product—lived experiences do not need to be perfect; if the connection between the person and their creation is there, the story is perfect.

Image 13.4 *No. 1. Where can I go?* Digital Sketches using Procreate on Apple iPad (Makayla Lewis, 2022 and 2023)

Image 13.5 *No. 2. Where can I go?* Digital Sketches using Procreate on Apple iPad (Makayla Lewis, 2022 and 2023)

Images 13.4 & 13.5: Where Can I Go?

A collection of reflective digital sketches that explore the meaning of HCI and art. Image 13.4: A self-portrait of Makayla holding up their creative outputs, trying to find a place to fit in HCI's closed, dark, somewhat more defined space (Sturdee, 2023). Top: Profile of Makayla looking, trying to find their way through the structured, yet scary, area of HCI (Lewis et al., 2023b). Bottom: An abstract representation of the tools available for creativity, suggesting a potentiality of engagement or lack thereof (Lewis, 2022).

Image 13.6: Technology and Me (or Me and Technology)?

A collection of digital sketches exploring my experiences, feelings, and attitudes towards technology. Top: Reflection in action sketch from 1st International Workshop on Explainable AI for the Arts (XAIxArts) at ACM Creativity and Cognition (C&C) 2023, where Makayla inserts her experiences, feelings and overall reflections about

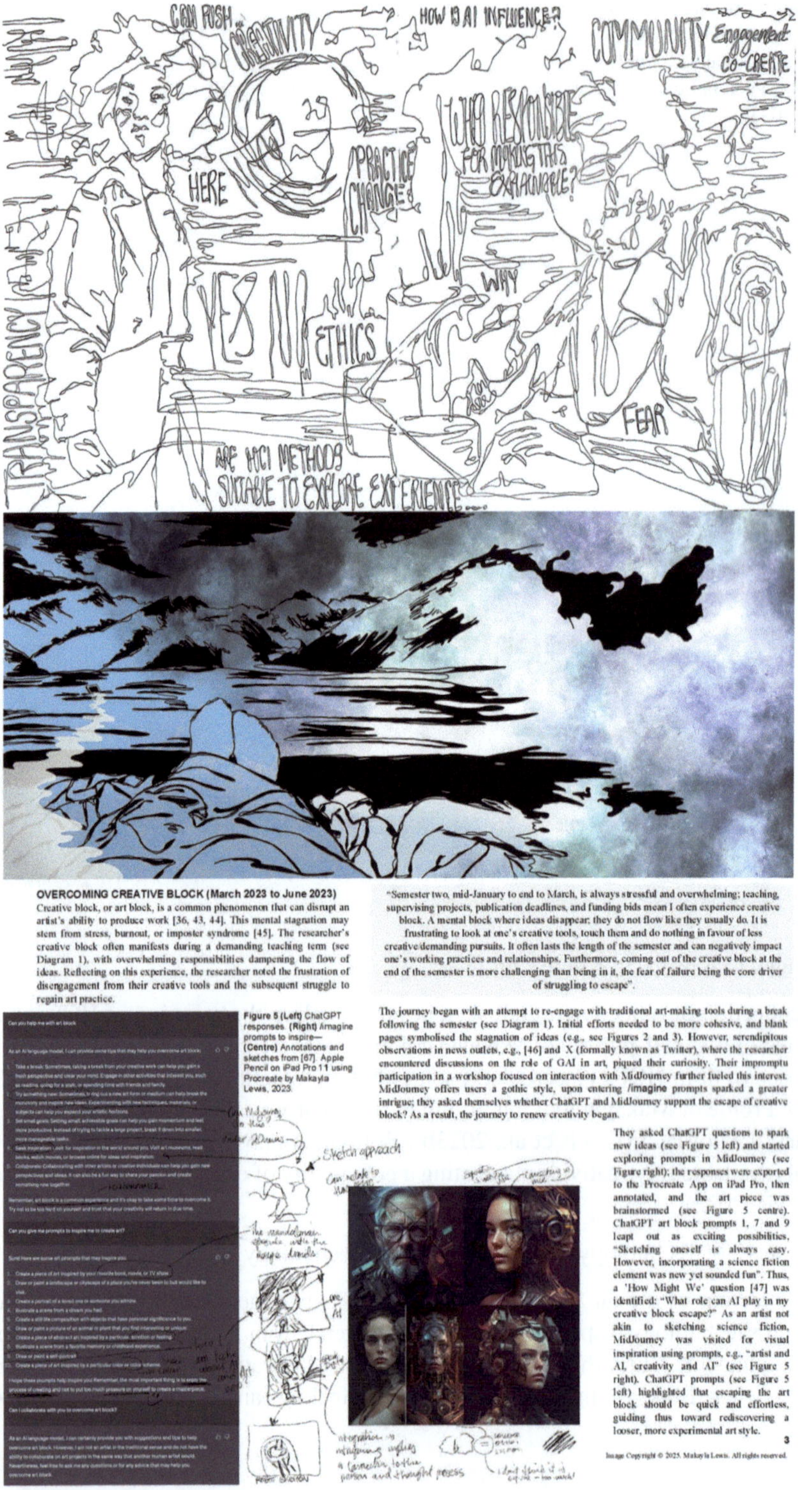

Image 13.6 *Technology and me (or me and technology)?* Digital Sketches using Procreate on Apple iPad (Makayla Lewis, 2022 and 2023)

the role AI is (and could) play in arts—an accumulation of their and the audience's discussions. Middle: Extract from 'Do Humans Dream of Digital Devices' (Sturdee, 2022); Makayla lays in bed looking at a mountainous unknown region with a vast lake between her feet and the shore on the other side. It is unclear why she is; it appears to be the vacuum of space, with a bright and colour nebula to the right. In the far distance is a smartphone sinking into the lake, a trail of light, maybe the journey it has taken, is connected to the phone and its body, maybe her hand. It is still being determined whether it is day or night. Bottom: Extract from 'AIxArtist' (Lewis, 2023) self-portrait in response to *ChatGPT* and *Midjourney* inspirational outputs 'AI and Me, 2023 "Maybe beyond?".

Image 13.7 & 13.8: Sharing Lived Experiences

A collection of digital sketches depicting Makayla's everyday experiences. Top: Sharing what it's like to experience an Absence Seizure.[2] Bottom: The potentiality of creating alongside other artists, an extract from 'Portraying What's in Front of You' (Lewis, et al., 2022b). Image 13.8: I am taking back my sketches.

13.5 From There to Here & Tarot and Technology

It is necessary to negotiate the distance one occasionally has with one creative practice by asking what purpose it should serve (Lewis et al., 2023c). Following the publication of "Plurality of Practices", I began creating space between my research and creative practices. Until that point, the two processes had been intimately intertwined (Gong et al., 2018; Davis, 2019; Davis & Wentzel, 2021; Davis et al., 2021), but putting space between them allowed my creative practice to become a refuge when the pressures of research and the perils of academia became overwhelming. Both of my practices thrived in this breathing room. Deeply inspired by Suzanne Treister's "Hexen 2.0", I became interested in the relationship between technology and the occult, how interacting with images becomes a technology of ideas, and how interaction produces meaning. This led me circuitously to tarot as an illustrative medium. Ironically, after investing significant time into illustrating "The Deleted World Tarot", researchers within the HCI community became interested in how Tarot can serve as a technology, and my creative project once again returned to academic work. Below, I will recount a more personal exploration of my interpretation of tarot illustration and how various tools I had previously developed for HCI academic, or creative coding venues were employed to render some of the image material.

The rise of technology and the allure of the occult share a common thread: the promise of a better world that never takes material form. Many queer people were raised in religious homes and believed in the 'utopian' vision promised by traditional spiritual teachings. However, we are often outcasts from these exact systems that

[2] https://www.epilepsy.com/what-is-epilepsy/seizure-types/absence-seizures#What-is-an-absence-seizure.

Image 13.7 *No. 1. Sharing lived experiences.* Digital Sketches using Procreate on Apple iPad (Makayla Lewis, 2022 and 2023)

Image 13.8 *No. 2. Sharing lived experiences.* Digital Sketches using Procreate on Apple iPad, 2022 and 2023

promise us utopia. Unsurprisingly, the history of tarot is filled with queer illustrators and practitioners. Pamela Coleman Smith's iconic Illustration for Rider and Waite remains one of the most prevalent collections of tarot images today. Much like the utopic visions of traditional religion, technologies of the past held similar promises of utopia. The early years of the Internet they offered unlimited knowledge and connection, an unprecedented era of enlightenment and the free exchange of ideas. These visions of better worlds also faded into the diss-information age of cyber superstitions and data-driven oppression in which we currently live. In both failed utopian visions, the one promised by traditional faiths and the one promised by emerging technologies, something was lost along the way. A choice was made. A world was deleted. The Deleted World Tarot is a tool for healing the rift between the choices of the past and the consequential loss of alternative futures. What is not saved will be lost.

Image 13.9: Calliope

My previous research focused on creating tools to foster creativity in collaborative efforts involving human artists and artificial intelligence, as detailed in (Davis et al.,

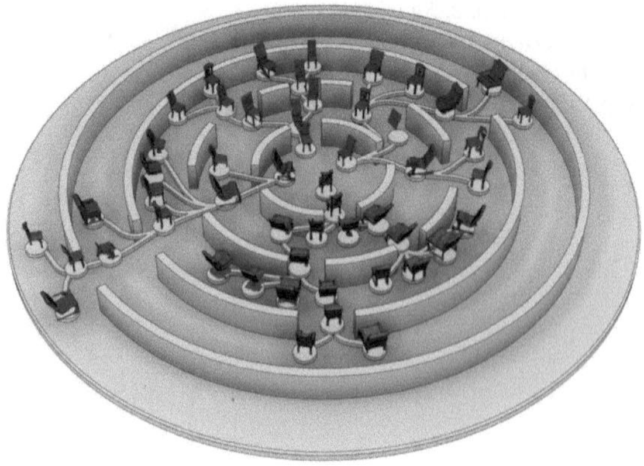

Image 13.9 *Calliope* used a labyrinth-like structure to enable users to collaborate with generative AI in virtual reality to produce design and creative 3D artefacts (Davis, 2021)

2021). This undertaking intricately intertwined my pursuits in both creative and engineering research domains.

Image 13.10: Hideout Colony Collapse

This drawing was created during the COVID-19 lockdown as a part of ongoing experiments blending human and machine contributions in image creation. 'Hideout Colony Collapse' was part of a series of drawings called "Hideouts", which visually represented the isolation of the pandemic lockdown and were a product of engineering experiments intended to provide the experience of creative collaboration with a machine since collaboration with other humans was impossible during lockdown.

Image 13.11: Tarot Cards

A curated collection of excerpts from the Deleted World Tarot demonstrating how creative coding and creativity support tool research can be transmuted into the analogue, physical tradition of tarot.

Image 13.12: The Deleted World Tarot

This collection depicts images rendered using traditional techniques but informed by concepts derived from engineering research.

13.6 The Threads Hold Thoughts in Place

In this chapter, I use textiles thinking to explore experiences and thoughts related to developing new project ideas. Below, I present pictures of a mixed-media embroidery hoop I made as I was developing a new project in the summer of 2023, followed by a piece of reflective writing that should be read alongside the pictures. Both the embroidery hoop and the text hold steady a collection of moments from the summer

Image 13.10 *Hideout Colony Collapse* hand and algorithmic ink drawing on paper, 2020

as I was developing new research ideas after completing a two-year collaborative research project.

In my academic work, few things are as exciting as developing new project ideas—the spirals of thought, of writing and editing, and editing, and editing, and editing, and editing.

Sometimes it all happens at once. Writing. Thinking. Making.

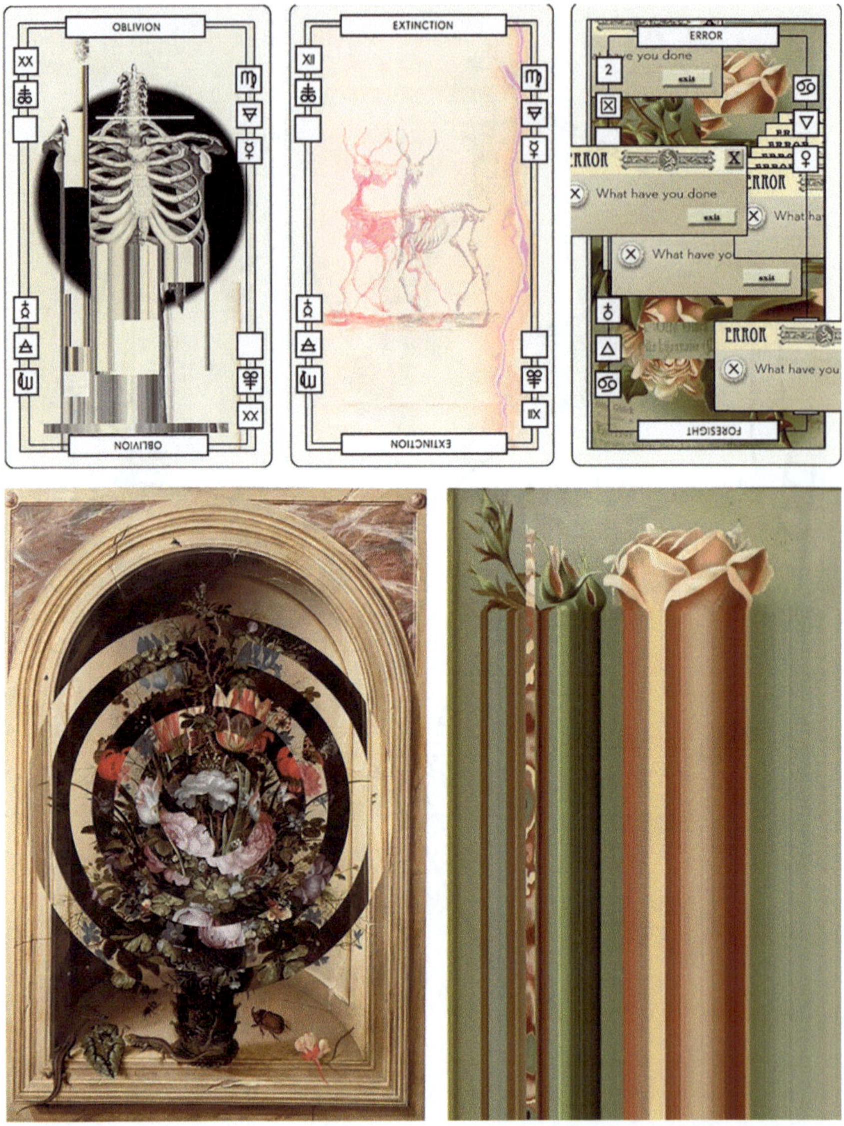

Image 13.11 A collection of images demonstrating how images generated through creative coding practices and HCI research are transmuted into designs for *tarot cards*, 2021–2023

It all starts with the comfort of the spiral—nothing new is needed. We don't need to reinvent the wheel. Instead, I learn about what is out there and what I have learnt from past projects, and I see where the current project is moving towards.

It starts with comfort. The spiral. Pinks and purples and specks of green. It starts with what I know—laying the groundwork, setting the scenes, and covering the

Image 13.12 A collection of tarot cards encapsulating the influence of recent research concepts on the practice of developing designs for *The Deleted World Tarot*, 2023

fabric with the base coat of stitches. At this point, I'm running, thoughts going fast, colours being put down quickly. Bulletpointed outlines of ideas. Many of them. All at once. Sometimes, work on two or three documents to put down thoughts. Typing. Sketching. Handwritten notes. Mind maps. And messy handwriting on too-big-pieces of paper for the thoughts.

The groundwork is messy. It's a complex web of threads, colours, and thoughts. I live in the mess of it. I stitch one thread of colour at a time, but looking at the combination of colours, what happens if I cross this thought with that? What happens if we read disconnected texts, bring them together, and layer them on top of one another? What if we cut them up, turn them around, and stitch into them?

I like this chaos. I like the mess. Both my hoop and my written documents are a mess. But that's where my thoughts seem to thrive—the mess of research, the mess of funding applications, the mess of unplanned stitches.

This isn't how it's supposed to be. The back of the hoops should be as beautiful as the front. Funding applications should be clear and innovative. It's not about mess. It's not about maintenance. But that's where my head goes. What is the mess? How can we complicate rather than simplify? How can we aspire to maintain hope, care, and support? It's all innovation. Change. But they don't want to hear the answers we have. That we know work. Do we really need to reinvent the wheel? Again.

It starts with the spiral. A messy web of colours. Pink. Purple. Green. I let things sit.

I rifle through my box of threads and a jar of cut-offs as I look at my bookshelf and rifle through the pockets of my brain that contain elements of papers and books I've read. I read the spiral as I read the paper. Just like concepts and practices, quotes from collaborators and participants, some threads jump out more than others.

It starts with a trickle of colour. But then I let inhibitions go and pick up the sparkly gold thread. What is the gold thread that goes through my work? A crisis of identity follows suit; like a spiral, it comes and goes, deepens, and brightens as it evolves.

The threads hold thoughts in place.

Soon, things become more messy. Specks of colour and ideas. Moments of clarity and moments of mess. I let myself follow the glimmer, the glow, the sparkle. Strengthened by the courage of others to allow for mess, to follow the glow, and to situate the research within an ongoing spiral, a programme of work. I don't need to reinvent the wheel.

And so I stitch on, picking colours spontaneously. Following what makes sense at the moment rather than what looks right. One stitch after the other. One word after the other. Exploring opportunities, creating textures of thought. Not shying away from the sparkle. Allowing for moments of excitement, of love, of red hearts and yellow stars. Allowing myself to get excited, to enjoy the process.

X marks the spot.

I pause.

I let it sit.

Colours and concepts overlap, tie together, wrap around one another, knot together, or simply lay next to one another as I create space, leave things blank, cover up previously visible parts of the spiral.

Don't focus on the outcome, I tell myself. It will make itself clear as the work progresses.

Editing. Stitching.

Editing. Re-writing.

Editing. Layering textures and colours and experiences and practices and thoughts.

Image 13.13 *No. 1 and 2. One stitch after the other.* A photograph was taken throughout the making of a mixed-media embroidery hoop to explore and capture experiences and thoughts related to 'developing new project ideas' as part of the research process. Thread and beads on Canvas (Angelika Strohmayer, 2023)

I'll know when it's done, when it's done.
(or when the deadline arrives).

Images 13.13, 13.14 and 13.15: One Stitch After the Other

One word after the other. Exploring opportunities, creating textures of thought, not shying away from the sparkle, allowing for moments of excitement, of love, of shiny plastic-y red hearts and yellow stars, and allowing myself to get excited and enjoy the process. X marks the spot. I pause. I let it sit. Colours and concepts overlap, tie together, wrap around one another, or lay next to one another as I create space, leave things blank, and cover up previously visible parts of the spiral. Don't focus on the outcome, I tell myself. It will make itself clear as the work progresses.

Editing. Stitching.
Editing. Re-writing.
Editing.
Layering textures and colours and experiences and practices and thoughts.
I'll know when it's done when it's done.
(or when the deadline arrives).

Image 13.14 *No. 3. One stitch after the other*. A photograph was taken throughout the making of a mixed-media embroidery hoop to explore and capture experiences and thoughts related to 'developing new project ideas' as part of the research process. Thread and beads on Canvas (Angelika Strohmayer, 2023)

Image 13.15 *No. 4. One stitch after the other,* pictures taken throughout the making of a mixed-media embroidery hoop to explore and capture experiences and thoughts related to 'developing new project ideas' as part of the research process. Thread and beads on Canvas (Angelika Strohmayer, 2023)

13.7 Refocusing My Work on Dancing First, Innovating Second

Since we last wrote the pictorial on Plurality of Practices, my artistic practice has not changed. However, my position with regard to what we develop in HCI has moved slightly, perhaps more radically. Indeed, I am still entangling my practice of dance and choreography with my academic endeavours in HCI, exploring how technology intervenes and the boundaries between these two fields and their values, goals, and priorities. However, I shifted in the way in which I see what technology means in these spaces and the need for it to be called and used. I am more and more attracted to the idea of less is more, and that sometimes simple setup where very few interactions are designed and made visible can yield the most intriguing and creative outcomes. For that, there are concepts in HCI that come in handy, such as removal (Homeward et al., 2020; Pierce, 2012). I would say I am becoming more and more low-tech and focusing more on how the studio space provides experimental ground for embodied work as such, mediated or not by technology. In the images I provide from the past, I put in RCO my piece that was heavily mediated by mobile phones (Fdili Alaoui & Matos, 2021).

Images 13.16, 13.17, 13.18 and 13.19: Performance

Image 13.16 A photograph of *Radical Choreographic Object interactive*, participatory dance piece where participants respond to the instruction on the phone by lying on the floor. Piece choreographed by Sarah Fdili Alaoui and Jean Marc Matos in 2018 copyright @Fabien Leprieult

Image 13.17 A photograph of the *CO/DA interactive dance improvisation* where Sarah Fdili Alaoui improvises with sensors worn on her body and a Jules Françoise live codes the sound-movement interaction on the fly. The improvisational piece made by Jules Françoise, Yves Candau and Sarah Fdili Alaoui in 2021 copyright © Christian MOREL / LISN / CNRS Photothèque

Image 13.18 A photograph of a scene from the dance performance *Vintage* choreographed by Sarah Fdili Alaoui and Annika Kompart in 2019. Copyright @ Performance Art Platform Aarhus

Image 13.19 A photograph of the donation booth for the *Vintage* performance made by Sarah Fdili Alaoui and Annika Kompart in 2019, where Sarah Fdili Alaoui is talking to 2 audience members and dancer Aline Sanchez is sharing a dance with two audience members inside the booth Copyright @ Performance Art Platform Aarhus

In the images I provide from the present, I put in Vintage, a piece I choreographed as both a dance artist and an HCI and interaction designer researcher and the de-centres the use of technology. I collaborated with another choreographer to create the performance. Our intention was to explore the idea of recycling dance. To do so, we (1) ran a series of rehearsals where we shared dances among the two of us, the choreographers, and an additional dancer, (2) participated in workshops with two communities of dancers that answered a call for participation, a group of Somali dancers and two Indian classical Bharatanatyam dancers, where they shared their dances with us and (3) organised two donation booths where participants donated their personal dances to us. To facilitate the collection of the dances to be reused, I designed a web application that allows dancers to capture video and sound of their dances. In the application, they can edit the audio-visual materials. They can also record a voice-over on top of it and add meta-data and explanatory text onto each item captured.

The choreographers, the dancers, and the participants in the donation booth used the application to create short documentaries of their dances, which helped us, the choreographers, learn them and reuse them as embodied choreographic material. This piece is different from my previous ones. It focuses on how our creative work was intimate and social, situated, and emergent, with unique relationships with communities of practice that shaped it entirely. The technology in Vintage is put in the background.

Indeed, the web application needs to provide new ways of recording dance or new interactions on screen. I can even claim it is an old-school screen-based, button-based user interface. I designed a humble technology that was meant to solely serve its purpose. I did not fantasise about it as something that would augment our practice or enable our creativity to unfold. Instead, I proposed a simple but robust, usable, and functional way of augmenting video capture and editing. I opted for simplicity over innovation. I wanted the artists and communities of practice to handle the creativity part and to be in the foreground of the process and the outcome.

13.8 Materials Explorations and Experiences within HCI and Beyond

In my ongoing research, I work with digital interactive technologies, exploring their role in crafting emotionally significant art digital jewellery (Koulidou & Mitchell, 2021). Inspired by contemporary jewellery practices and my fascination with found objects, I designed an autobiographical piece, "Thalassa" (2021–22), connecting the wearer with the essence of the sea and its dynamic state. Digital jewellery, in my practice, acts as a catalyst for connections between people, places, and times. Since our last publication, I took a senior lecturer position in Art and design at Sheffield Hallam University, leading the BA Jewellery, Materials, and Design Course. Immersed in specialised workshops for silversmithing, ceramics, welding, and 3D printing facilities, I discovered a rich source of inspiration through experimentation and dialogues with creative practitioners. My creative practice clearly centres on material explorations and experiences [Karana et al., 2015], influenced by the vibrant artistic culture at Sheffield Hallam.

While interactions with HCI and interaction design colleagues persist, my current institution's emphasis on art and design transforms the dynamics of conversations and opens avenues for different collaborations and explorations. The tension between my HCI/artistic roots and my evolving creative practice becomes apparent as I engage more with craft practices. Originally, my making experiments aimed to enhance my digital jewellery practice, from designing cases for electronics to exploring porcelain as light diffusers. Surprisingly, perceived "failures" like a mould for slip-casting that traps air bubbles and a clogged 3D printer nozzle birthed new artistic avenues. What surprised me most was the emergence of artistic explorations detached from my digital jewellery research, such as the transformation of tree twigs into a standalone art piece (currently under development). The realisation that my making could exist independently of my research constraints sparked a shift in my approach. Making, synonymous with thinking and sense-making [Ingold, 2013], found liberation within the established methodology of an Art and design department. This freedom steered my research towards a broader perspective, envisioning future research and creative directions that seem to be moving away from the HCI community. Currently, I'm working towards a resubmission for an HCI conference and a revised HCI journal paper to encapsulate the wealth of research conducted with my students on digital art jewellery (2021–23). I am considering other venues where my research can be published and exhibited. I am also exploring the role of jewellery practices in fostering cross-cultural connections within migrant communities in Sheffield as an artist and volunteer in a local organisation. Whilst I continuously explore materials and new craft practices. The tension between HCI and art has dissipated in my work, paving the way for a more expansive and inclusive research direction that sometimes is separated from my creative practice and, at other times, is an integral part of my making.

Image 13.20: Material Explorations and Experiences #1

Thalassa (2021–22) by Nantia Koulidou. A necklace that connects the wearer with the sea. 80 mm × 50 mm x 25 mm. Material explorations: found tin, steel, fish line, leather, and electronics (Adafruit Feather huzzah32, LiPo Battery, NeoPixel Flora 2).

Image 13.21: Material Explorations and Experiences #2

Explorations of light diffusers using porcelain, porcelain paper clay, washi paper and traditional jewellery-making techniques such as prong setting. Design development for a pair of connected digital brooches.

Image 13.22: Material Explorations and Experiences #3

Design development of a digital piece that connects the owner with nature and hiking. The piece was initially designed to incorporate electronics. However, challenges in the making and opportunities that arose from the process changed the development.

Image 13.23: Material Explorations and Experiences #4

Collaborating with Ultimaker Material Station printer to design a digital piece. The piece was initially designed to be printed with one filament (wood); however, during the printing, the nozzle clogged, and the printer started printing with another filament preloaded (white PLA) to complete the task. Images show the development of the final prototype made from found objects, wood and white PLA filament and electronics.

13.9 Meandering Materials

Last time, I wrote about and drew on theories I was exploring for other work to make sense of theories. It was an exploratory and material process, one I found reminiscent of tendrils exploring knowledge through visualising it in ways that were not relying on words but rather on concepts. They didn't need to make sense to anyone; they needed to help me understand. These days, I seem to still draw analytically, even predominantly so. However, it has moved to a digital space, one that is more literal and less open, but then, what I analyse these days are not concepts but materials and interactive setups. So, in this regard, it might make sense that my drawing shifted

Image 13.20 *Material Explorations and Experiences #1* (Nantia Koulidou, 2023)

Image 13.21 *Material Explorations and Experiences #2* (Nantia Koulidou, 2023)

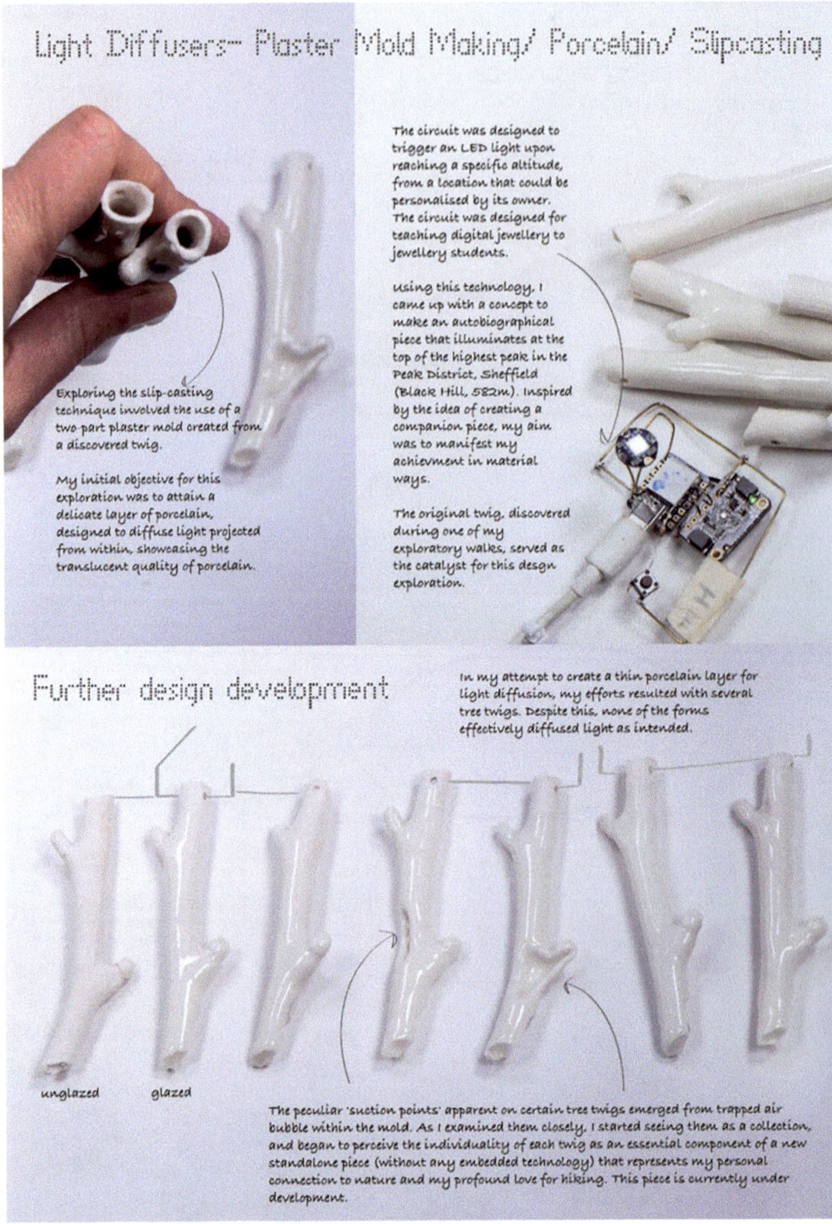

Image 13.22 *Material Explorations and Experiences #3* (Nantia Koulidou, 2023)

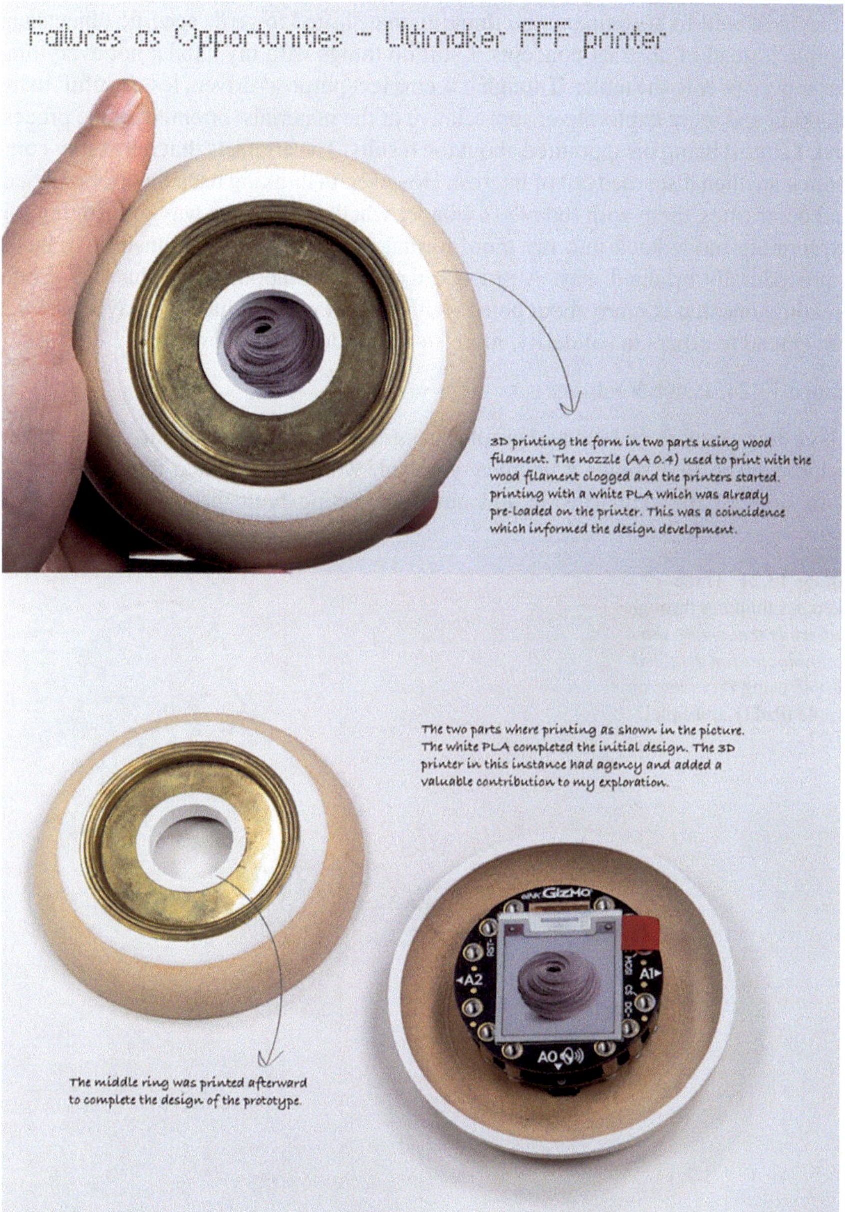

Image 13.23 *Material Explorations and Experiences #4* (Nantia Koulidou, 2023)

in style as well to accompany the thinking that shifted towards specific objects and people instead of abstract concepts. I still do things with my hands; not everything has moved over to the tablet. Though it seems less purpose-driven, less 'useful' to my thinking and more explorative, appreciative of the materials, oriented on the process and, I almost being disappointed about the results. The artefacts that inevitably come from it are then discarded out of interest. However, in thinking back on prior practices and contrasting them with today's, I wonder whether there is a way of bringing this exploratory mode back into my more formal thinking, less in an analytical, but in a procedurally oriented way. A practice that is more about doing than it is about creating, one that is more about being (with) than it is about having, maybe one that can extend to others in solidarity, more so than it does right now.

Image 13.24: Harsh Realities

Three-line sketches on blue and purple backgrounds on harsh disability-related technologies. The first shows a cochlear implant, which is a device that works from both sides of the skull (inside and out), connecting both through magnetism. It

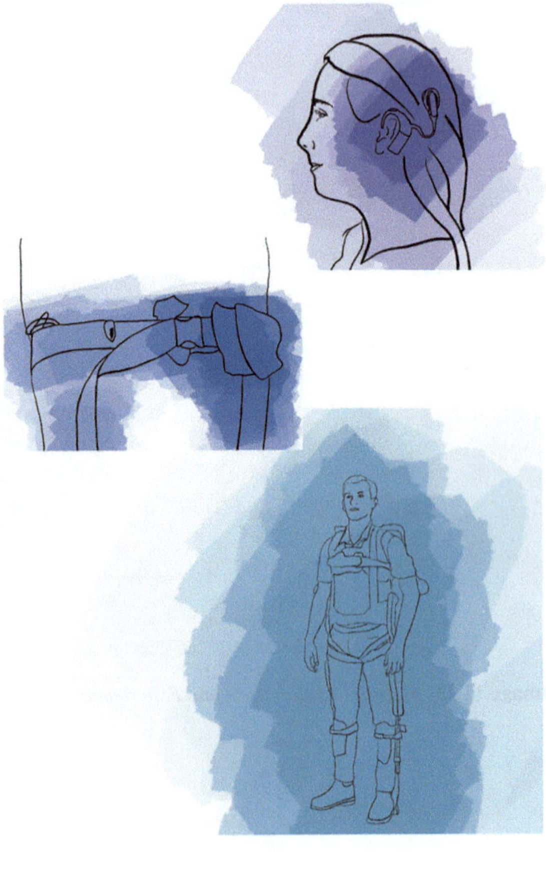

Image 13.24 Three sketches thinking through *materials that shape the technologies for disabled people* using Procreate on Apple iPad (Katta Spiel, 2023)

provides substandard hearing to a subgroup of deaf people but is often already implanted in non-consenting children and particularly pushed on hearing parents of deaf children (Sparrow, 2005). The second image shows a belt supposedly keeping kids with ADHD from blurting (Smit & Saskia Bakker, 2015). It has strong materials that particularly tie the body together. The third image illustrates a person wearing an exoskeleton. It binds their entire body, supporting legs and upper body simultaneously.

Image 13.25: Ambivalent Relationships

Three-line sketches on coloured backgrounds on disability-related technologies that make room for ambivalence. The first illustrates a young person wearing smart glasses with a blue-tinted background. Since smart glasses can be seen as a platform device with multiple options (and not just the limited ones that technologists imagine, most prominently, for autistic children (Washington et al., 2017)), they bring with them some amount of self-determined interaction with the device itself. The second image shows a person using their smartphone to check their blood glucose levels from a sensor attached to their upper arm with a salmon-coloured background. These are

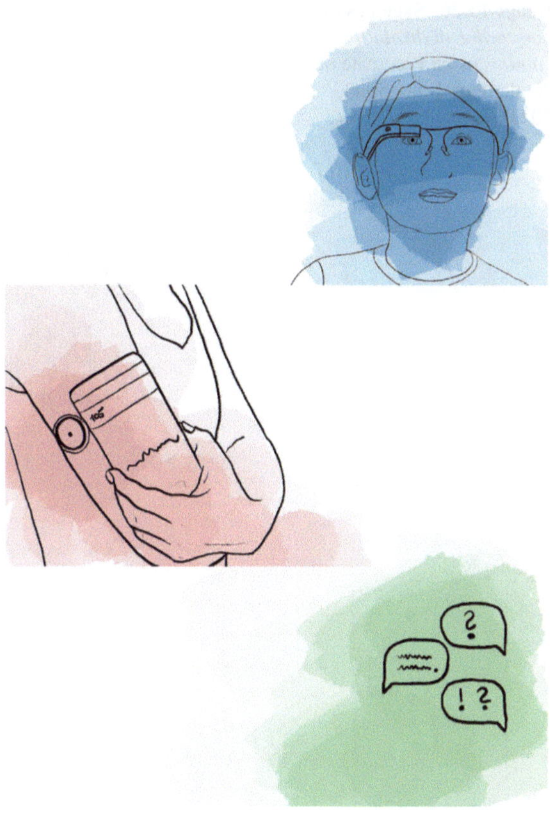

Image 13.25 Sketches of *technologies disabled people use* with varying shades of self-determination using Procreate on Apple iPad (Katta Spiel, 2023)

often praised, even by diabetics, for providing detailed insights and aid in managing diabetes well. However, they are also commercialised body networks (Pedersen & Iliadis, 2020). The third image shows chat bubbles on a light green background as a stand-in for the mutual support networks disabled people create for themselves (Piepzna-Samarasinha, 2018). These technologies, though, have specific in-and-out mechanisms and are often criticised for being inaccessible in some regard or another.

Image 13.26: Co-constructing Communities

Four-line sketches on pink backgrounds detailing different aspects of disabled embodiment and the cultural implications thereof. The first shows a person with long curly hair balancing a pencil on their upper lip, looking into the air. This illustrates the notion that people with ADHD often need something to fidget with to concentrate (Graziano, 2020). The second image shows a child pointing at toy cars lined up in a row. This is in reference to autistic people exhibiting different kinds of socialities (and play) when engaging with others despite them often remaining unacknowledged as equally valid to those of neurotypical people (Milton, 2014). The third image depicts a person using a wheelchair being pensive but determined.

Image 13.26 Sketches on *non-technical disability cultures* using Procreate on Apple iPad (Katta Spiel, 2023)

The last image shows a deaf person, specifically Nyle DiMarco, signing in American Sign Language.

13.10 Summary

We present seven perspectives on a plurality of practice, tension between art and research, resolved or unresolved. The images presented here are a curated selection created in the intervening years since the pandemic in 2020, and the world has changed again. As we consider our personalities and achievements, we must also place our work against the backdrop of change and the emergence of digital genres, which simultaneously pull apart and reform the world around us. To create makes us human; to present our works to others suggests we are artists. Does writing about our creations make us researchers? This question can remain unanswered for now. It is enough to be.

> We are not only researchers. We are artists, makers, dancers, sculptors. We connect with people, inspire, indulge and create. We find ways of expression in parallel to our identity as researchers, and sometimes, these expressions find their way into our working world.

Acknowledgements Makayla would like to acknowledge that the outputs and prompts provided by ChatGPT and MidJourney in Image 13.6 (Bottom) are not owned by the author or reflective of the views of the product designers, developers, or service providers. Sarah would like to acknowledge Jules Françoise, Yves Candau, Jean-Marc Matos, Annika Kompart and Aline Sanchez, as well as all the dancers and audience members who participated in the performances RCO, CO/DA and Vintage that she referred to. Katta would like to acknowledge their disabled, crip and deaf peers, who continue to be a tremendous inspiration, a source of solidarity, and a force to be reckoned with. Nantia would like to acknowledge all the technicians and colleagues at Sheffield Hallam for their continuous support and encouragement to use making as a way of thinking. Angelika would like to acknowledge her collaborators and colleagues for doing important work and encouraging engagement with mess and material exploration. Miriam would like to thank all those above who continue collaborating within this space, especially Makayla Lewis, who continues to inspire and push her to remember her artistic side.

References

Alaoui, S. F., & Matos, J.-M. (2021). RCO: Investigating Social and Technological Constraints through Interactive Dance. In *Proceedings of the ACM Conference on Human Factors in Computing Systems (CHI) online.*

Blair, K., Sturdee, M., Vermeulen, L. M., & Oehlberg, L. (2022). Art is Not Research. Research is not Art. In *CHI Conference on Human Factors in Computing Systems Extended Abstracts* (pp. 1–9).

Davis, J. U., & Wentzel, J. (2021). Font Your Friends and Loved Ones: On the Utility of Ugly Interfaces. In *Extended Abstracts of the 2021 CHI Conference on Human Factors in Computing Systems (CHI EA '21).* Association for Computing Machinery, New York, NY, USA, Article 28, 1–7.

Davis, J. U., Anderson, F., Stroetzel, M., Grossman, T., & Fitzmaurice, G. (2021). Designing Co-Creative AI for Virtual Environments. In *Proceedings of the 13th Conference on Creativity and Cognition (C&C '21)*. Association for Computing Machinery, New York, NY, USA, Article 26, 1–11.

Davis, J. U. (2019). IllumiWear: A Fiber-Optic eTextile for MultiMedia Interactions. In *Proceedings of the International Conference on New Interfaces for Musical Expression.*

Ellis, C., Adams, T. E., & Bochner, A. P. (2011). Autoethnography: An Overview. Historical, Social Research. *Historische sozialforschung, 36*(4), 273–290

Goldschmidt, G. (2008). Visualization-Sketching is alive and well in this digital age. In *Design Processes* (pp. 28–43). IOS Pres

Gong, J., Yang, X., Seyed, T., Davis, J. U., & Yang, X.-D. (2018). Indutivo: Contact-Based, Object-Driven Interactions with Inductive Sensing. In *Proceedings of the 31st Annual ACM Symposium on User Interface Software and Technology (UIST '18)*. Association for Computing Machinery, New York, NY, USA, 321–333.

Graziano, P. A., Garcia, A. M., & Landis, T. D. (2020). To fidget or not to fidget, that is the question: A systematic classroom evaluation of fidget spinners among young children with ADHD. *Journal of Attention Disorders, 24*(1), 163–171.

Homewood, S., Karlsson, A., & Vallg.rda, A. (2020). Removal as a Method: A Fourth Wave HCI Approach to Understanding the Experience of Self-Tracking. In *Proceedings of the 2020 ACM Designing Interactive Systems Conference (Eindhoven, Netherlands) (DIS '20)*. Association for Computing Machinery, New York, NY, USA, 13.

Ingold, T. (2013). *Making: Anthropology, Archaeology, Art and Architecture*. Routledge.

Karana, E., Pedgley, O., & Rognoli, V. (2015). On Materials Experience. *Design Issues, 31*(3), 16–27.

Kimbell, L., Carlebach, E., Smyth-Allen, H., Gherhes, C., Lewis, M., & Vorley, T. (2021). *AI Readiness: A Collaborative Design Toolkit for Professional Service Firms*. Oxford Brookes University.

Koulidou, N., & Mitchell, R. (2021). Art Digital Jewellery: Practitioners' Perspectives. In *Proceedings of the Fifteenth International Conference on Tangible, Embedded, and Embodied Interaction (TEI '21)*. Association for Computing Machinery, New York, NY, USA, Article 29, 1–11.

Lewis, M., & Coles-Kemp, L. (2014). A Tactile Visual Library to Support User Experience Storytelling. In *NordDesign2014.*

Lewis, M., Sturdee, M., Davis, J. U., Gamboa, M., Alaoui, S. F., Ohlenschlager, C. E., Gaver, W., Blevis, E., & Loke, L. (2023c). Is it Art, is it HCI? Exploring Tensions Between Practice and Research. In *Extended Abstracts of the 2023 CHI Conference on Human Factors in Computing Systems (CHI EA '23)*. Association for Computing Machinery, Article 525, 1–4.

Lewis, M., Sturdee, M., Miers, J., Davis, J. U., & Hoang, T. (2022a). Exploring AltNarrative in HCI Imagery and Comics. *AltChi Paper to 2022 CHI Conference on Human Factors in Computing Systems*. Association for Computing Machinery, New York, NY, USA, Article 164.

Lewis, M., Toselli, M., Baker, R. M., Rédei, J., Ohlenschlager, C. E. (2022b). Portraying What's in Front of You: Virtual Tours and Online Whiteboards to Facilitate Art Practice during the COVID-19 Pandemic. In *Creativity and Cognition (C&C '22)*. Association for Computing Machinery.

Lewis, M., Sturdee, M., Gamboa, M., & Lengyel, D. (2023a). Doodle Away: An Autoethnographic Exploration of Doodling as a Strategy for Self-Control Strength in Online Spaces. In *CHI Conference on Human Factors in Computing Systems (AltCHI), April 23–28, 2023, Hamburg, Germany*. ACM.

Lewis, M., Sturdee, M., Hoang, T., Alaoui, S. F., Strohmayer, A., & Samuelsson-Gamboa, M. (2023b). Futuring (CHI)Art: Building a Collective Common Future. In *CHI Conference on Human Factors in Computing Systems Extended Abstracts (CHI '23 Extended Abstracts), April 23–28, 2023, Hamburg, Germany*. ACM.

Lewis, M. (2022). UX-Design Tools Mindfulness: 'One-Stop Shop' for Learning and Practice. In *Proceedings of the ICHI '22 Workshop InContext: Futuring User-Experience Design Tools. April 29-May 5, 2022, New Orleans, LA, USA.*

Lewis, M. (2023). AIxArtist: A First-Person Tale of Interacting with Artificial Intelligence to Escape Creative Block. In *the 1st International Workshop on Explainable AI for the Arts (XAIxArts).* ACM Creativity and Cognition (C&C) 2023. Online.

Milton, D. E. M. (2014). Autistic Expertise: A Critical Reflection on the Production of Knowledge in Autism Studies. *Autism, 18*(7), 794–802.

Pedersen, I., & Iliadis, A. (Eds.). (2020). *Embodied Computing: Wearables, Implantables, Embeddables, Ingestibles.* MIT Press.

Piepzna-Samarasinha, L. L. (2018). *Care Work: Dreaming Disability Justice.* Arsenal Pulp Press.

Pierce, J. (2012). Undesigning Technology: Considering the Negation of Design by Design. In *Proceedings of the SIGCHI Conference on Human Factors in Computing Systems (Austin, Texas, USA)* (pp. 957–966) *(CHI '12).* ACM.

Smit, D., & Bakker, S. (2015). BlurtLine: A design exploration to support children with ADHD in classrooms. In *Human-Computer Interaction–INTERACT 2015: 15th IFIP TC 13 International Conference, Bamberg, Germany, September 14–18, 2015, Proceedings, Part IV 15* (pp. 456–460). Springer International Publishing.

Sparrow, R. J. (2005). Defending Deaf Culture: The Case of Cochlear Implants. *Journal of Political Philosophy, 13*(2), 135–152.

Sturdee, M., Lewis, M., Strohmayer, A., Spiel, K., Koulidou, N., Alaoui, S. F., & Urban Davis, J. (2021). A Plurality of Practices: Artistic Narratives in HCI Research. In *Creativity and Cognition (C&C '21).* Association for Computing Machinery, Article 35, 1.

Sturdee, M., Lewis, M., & Miers, J. (2022). Do Humans Dream of Digital Devices? Subconscious User Experiences and Narratives. In *Creativity and Cognition (C&C '22).* Association for Computing Machinery.

Sturdee, M., Lewis, M., & Gamboa, M. (Eds.). (2023). *The State of the (CHI)ART. figshare.* Book.

Washington, P., Voss, C., Kline, A., Haber, N., Daniels, J., Fazel, A., De, T., Feinstein, C., Winograd, T., & Wall, D. (2017). SuperpowerGlass: A Wearable Aid for the At-Home Therapy of Children with Autism. *Proceedings of the ACM on Interactive, Mobile, Wearable and Ubiquitous Technologies, 1*(3), 1–22.

Chapter 14
Looking Over the Details

Andy Boucher

Abstract This chapter revisits a photo essay, written over a decade ago, to re-examine how it documented the design process in a Research through Design project. *Attention to Detail: Annotations of a Design Process* was crafted to challenge the confines of traditional academic formats and to demonstrate the value of making practice-based research visible. This retrospective aims to enrich the original publication by shedding light on our intentions, offering a fresh look at previously unpublished photographs from the project, and providing candid insights into the project's execution. The chapter underscores the lasting significance of visually chronicling the design process, which can reveal the intricate and often obscured layers of development that are critical to design research.

14.1 Introduction

Looking back to 2012, pictorial tracks had not yet been developed for ACM conferences. As a team of practice-based researchers at the intersection of design and HCI, we were frustrated by the limited opportunities available at the time to demonstrate the intricacies of our design process in traditional publications. We believed that illustrating the process of our Research through Design practice could help demystify design and show how the designs themselves can embody research. By then, we had been working on a project that was focused on batch-producing what I would now call research products for multiple participants. This was our first attempt at low-volume manufacturing, and it forced us to take scale seriously as part of the design process (Boucher, 2023). In practice, that made the work intensive and highly iterative: we ended up producing far more variations of each detail than copies in the final batch. Our approach was more akin to a traditional craft process than might be expected in this context, and we felt compelled to share the detailed nature of this work. This is where the concept of a photo essay emerged as a provocation: a way to challenge our community to devise a better method for presenting design work.

A. Boucher (✉)
Interaction Research Studio, Northumbria University, London Campus, London, UK
e-mail: andy.boucher@northumbria.ac.uk

S. An Blevis et al. (eds.), *Seeing Interaction Design*, Design Research Foundations,
https://doi.org/10.1007/978-3-032-14200-9_14

The Indoor Weather Stations (Gaver et al., 2013) were designed to reveal the microclimate of a home through three devices that draw attention to subtle draughts, ambient light (colour and intensity), and temperature differentials. We produced twenty sets of these devices for participants living in and around the South London neighbourhoods near our campus. For our photo essay (Jarvis et al., 2012), we decided to document the process of producing these devices in a deliberately ordered way. We hoped that curating a visual story, built from carefully selected stages of our design process, would help reviewers see the value of the contribution and the decisions made during development. We decided to adopt a photographic style influenced by Taryn Simon's 'Contraband': a series of 1,075 photographed objects detained at JFK Airport for inspection. In this work, each contraband item was framed within a large white space to expose its finer details. We approached our photographic study in a similar way, allowing us to interrogate our own design process and expose the intricate details we considered at each stage.

Our photo essay featured 112 photographs capturing stages in the crafting of the Indoor Weather Stations. It was densely packed, and we encouraged readers to scrutinise the images not on the printed page, but via a Flickr link (http://tinyurl.com/attn2detail) that, to my surprise, still works today. We submitted the piece to NordiCHI 2012. I remember the reviewers' mixed reactions to our approach, and can only assume that it was resuscitated during the PC meeting by a lone voice who wanted it to be accepted (whoever that was, thank you). I also remember a few stinging comments after the conference presentation. "Like a parent cherishing their child's drawing, this work only has meaning to you," is one that I perversely enjoy.

Pictorials, of course, are now commonplace, and it is interesting to reflect on how this work might have been presented had that format been available at the time. At the time, we felt we had to show a lot to make the case that the contribution was valuable, which is why there are so many images. Now, I think that while there is strength in the depth and density, there many photographs that I wish had more space to breathe. While working on this chapter, I looked back through my archive and found many unused photographs. By presenting a selection here, I hope to give some details the space they lacked in the original format. I also want to reveal parts of the design process that sat outside our original photographic style.

14.2 Contributions

This chapter revisits *Attention to Detail* and extends it through a retrospective examination of the design work and the forms of detail constrained by the original format. The images presented here foreground aspects of making, iteration, and assembly that were present in the original project but largely invisible in the photo essay itself.

Image 14.1: The Indoor Weather Stations.

The finished set of devices. The Temperature Tape (left) is designed for people to explore temperature gradients within the home. Two 2.5-metre fabric ribbons

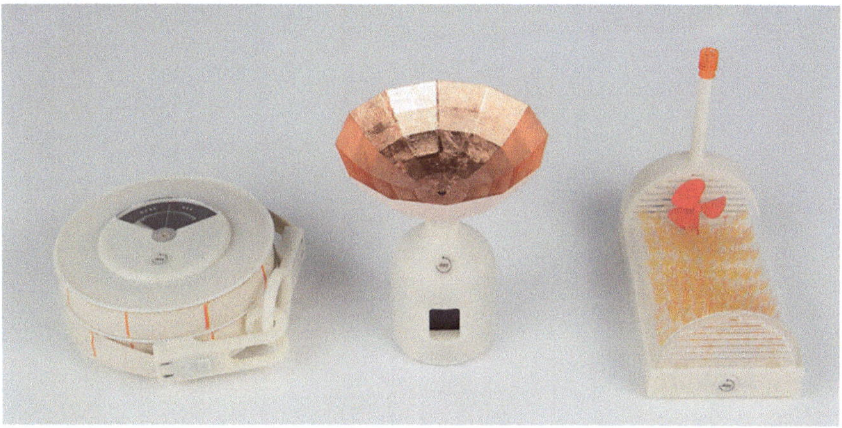

Image 14.1 The Indoor Weather Stations

connect to temperature sensors embedded at each end in small plastic bodies, one shaped like a hook and the other an eye. The central hub, similar to a wind-up tape measure, displays the temperature difference between the sensors on a needle gauge. The ribbon is screen-printed with thermochromic ink, changing colour between 15°C and 25°C to visualise temperature changes. The Light Collector (middle) captures and displays ambient light variations on a small OLED screen. It features a copper-lined funnel with an RGB sensor at its centre, measuring ambient light data that is transformed into a coloured horizontal pixel strip on the display. The Wind Tunnel (right) makes noticeable almost imperceptible air movements. It features a paper forest within a clear canopy over a 3D-printed landscape. Air currents are detected by a sensor in a chimney-like structure, which then activates a fan to simulate wind effects on the paper trees. A shared feature of all three devices is the 'last day' button, which enables playback of the last 24 hours of collected data.

Image 14.2: Examples of speculative sensor concepts, early experiments, and Cultural Probes.

Speculation and experimentation with domestic IoT sensors have long been a recurring thread in our design practice. A previous project (Gaver et al., 2009) led us to examine the legibility of sensors, hypothesising that this would result in sensor-based systems being more transparent to participants. This built upon earlier work, demonstrated here with a page from a 2001 workbook (top image), featuring sensor components from a speculative parts catalogue for a Home Health Monitor that users would assemble themselves. The bottom left image shows an analogue draught sensor constructed of tissue paper and jewellery wire. It formed part of a study considering mechanical extensions to electronic sensors to enhance the legibility of solid-state devices. The bottom right image shows an example of a cultural probe created for a domestic study we ran at the start of the project to prompt people to consider the microclimate of their homes. This probe asks participants to consider the contents of

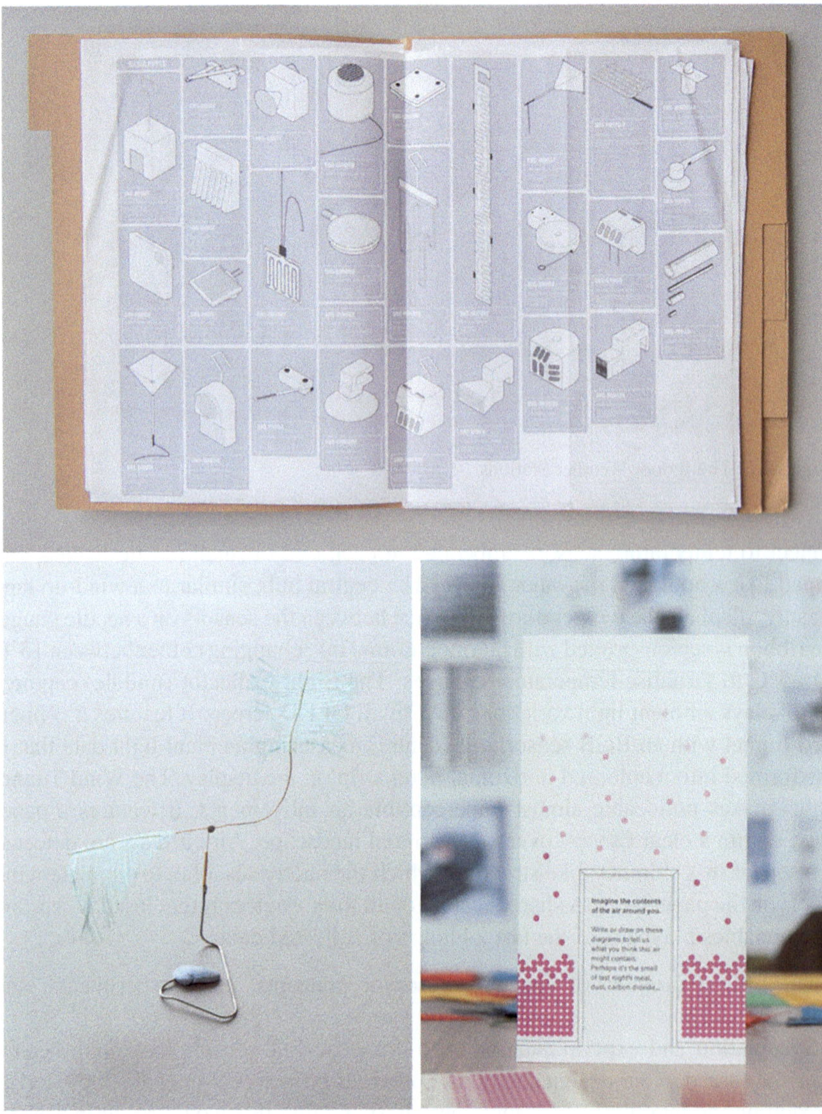

Image 14.2 Examples of speculative sensor concepts, early experiments, and Cultural Probes

air in their households, a task that often yielded unexpected results. For instance, one participant noted that their air included glue, oil, dust, mould, mouse killer, sugar, love, anger, flatmates, and insects, among other eclectic things.

Image 14.3: A collection of experiments with components, materials and forms.

For this project, we acquired a Stratasys Dimension Elite 3D printer, our first machine of this kind, with the intention of batch-producing the casings for the Indoor

Image 14.3 A collection of experiments with components, materials and forms

Weather Stations. Partly because we could, and partly as a learning process, we made many iterations of every component using this machine. We learned its quirks, discovered the best way to orient parts on the build plate for an optimal finish, and developed a reliable snap fit. We rapidly built an archive of development parts, often annotated with a marker pen to note faults, improvements, and general observations. These parts began to fill every available surface in our studio. We were also experimenting with a laser cutter. We etched Perspex with graphics and information for the Temperature Tape dial window, and optimised the slots in the end cover of the Wind Tunnel to maximise the buffeting of the paper forest.

Image 14.4: Multiple circuit boards inside the Light Collector.

The Indoor Weather Stations were constructed entirely using Microsoft's then-new Gadgeteer platform (Villar et al., 2010). The system supported rapid prototyping of embedded devices using electronic modules linked by fine ribbon cables. Although this modular hardware encouraged an exploratory design process with sensors, actuators, and displays, in retrospect it was not the most straightforward platform for batch-producing 60 finished devices. Pictured here is the Light Collector with its casing open. Despite its compact size, the device houses seven PCBs (circuit boards), each screwed into place within the enclosure and interconnected with numerous ribbon

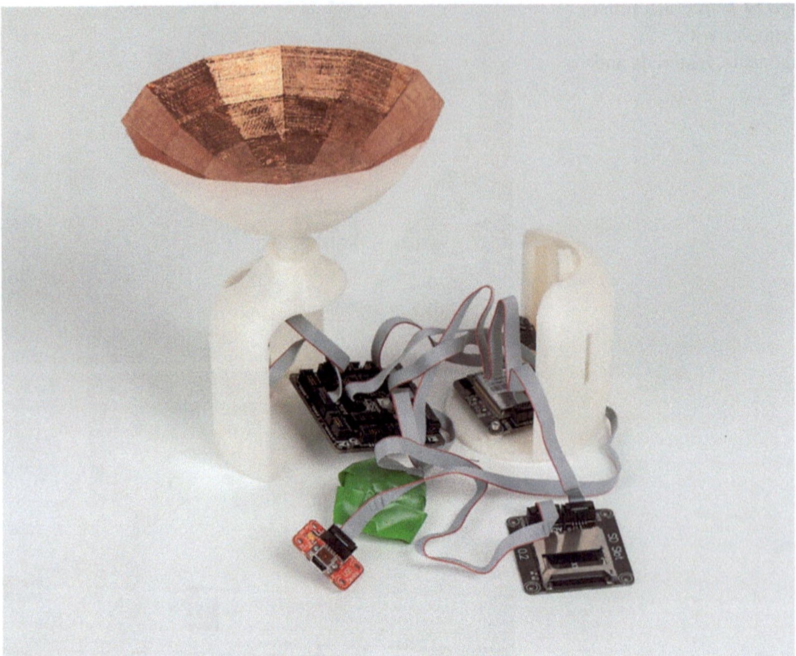

Image 14.4 Multiple circuit boards inside the Light Collector

cables. The assembly was intricate and demanded a significant amount of practice (and patience) to do proficiently.

Image 14.5: Testing the Light Collector and an early copper-leafing experiment.

Light Collector includes a funnel that can be positioned to direct the RGB light sensor at its centre towards a light source. The connection between the upper and lower parts is a 3D-printed ball-and-socket joint, allowing three-dimensional articulation without rotation that could strain the ribbon cable (an innovation that was briefly patented by our university's enthusiastic business team). We experimented with copper leafing on the funnel (bottom image), not to enhance the light-sensing capabilities but rather to refine the perceived finish of the 3D-printed parts, countering the somewhat undesirable ridged quality of the printed layers. The Light Collector's screen (top image) displays the sensed colour as a one-pixel-high horizontal strip. As it scrolls down one pixel every five minutes with each new sensor reading, it forms a segmented timeline of the last two hours of ambient light. The oldest readings sit at the bottom and the most recent at the top. Pictured here is a test of the sensor using coloured paper.

Image 14.6: A partially disassembled Wind Tunnel.

Like the Light Collector, the Wind Tunnel houses multiple PCBs, most arranged on a single plane beneath the paper forest and topographical base. Assembly is complicated by the need to mount and thread wires from the draught detector and

Image 14.5 Testing the Light Collector and an early copper-leafing experiment

Image 14.6 A partially disassembled Wind Tunnel

fan motor through the chimney housing and down to screw terminals on one of the PCBs. Fitting the printed and laser-cut paper forest into narrow slots in the 3D-printed landscape also requires precise work.

Image 14.7: Experimenting with the Wind Tunnel paper forest.

The final paper forests were crafted from separate layers of laser-cut vellum (plasticised paper), inserted into slots along the printed topographical base. Early experiments were mostly done in standard coloured paper, exploring different shapes, cut angles, and paper density. Each prototype was tried inside a testing mule of the tunnel and assessed for perceived turbulence and motion.We aimed for the forest to appear highly animated at high fan speeds, while still showing gentle ripples at low speeds. Perfecting this effect required numerous iterations, but the design eventually converged on a paper tree with a narrow trunk and multiple cut 'leaves' to catch the air. We switched from paper to vellum after discovering that paper structures lose elasticity over time and fail to return to their original shape.

Image 14.8: The Temperature Tape front, back, and partially unwound.

Like the other devices, the Temperature Tape houses numerous PCBs internally, as well as an analogue needle dial display. A distinctive feature, however, is the printed layers of thermochromic ink along the lengths of the fabric sleeves that encase the ribbon cable (top image). We used four different inks that become translucent at

Image 14.7 Experimenting with the Wind Tunnel paper forest

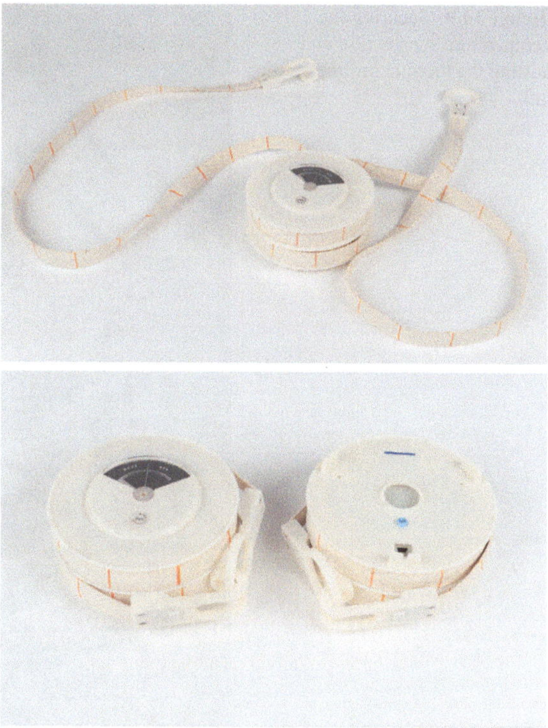

Image 14.8 The Temperature Tape front, back, and partially unwound

specific temperatures within a narrow range between 15°C and 25°C, causing the stripes to change from black below 15°C to red, then orange, and finally yellow around 25°C. These photos were taken on a mild day, with the orange stripe indicating a temperature of around 20°C. The final design for the rear casing of the Temperature Tape, which did not feature in the original photo essay, is shown here to highlight the large recess at the bottom of the housing. This recess allows the device to be balanced on and gripped by a thumb, making it easier to unwind the ribbons with the device in hand.

Image 14.9: Securing the temperature sensor case and testing the thermochromic ink.

During development, we frequently revisited components after flaws emerged through testing. The temperature sensor casings for the hook and eye were revised more than any other part. The hook, in particular, required several redesigns to prevent snapping caused by layer delamination. Through this process, we learned a great deal about 3D-printing and began to think about these forms in terms of layer construction, rather than as volumes agnostic of their fabrication. Another aspect that required multiple revisions was the method used to clamp the housing over the ribbon cable and fabric sleeve. Early versions made it easy to pull the ribbon cable out of the sensor PCB socket during use, sometimes tearing the socket from the PCB

Image 14.9 Securing the temperature sensor case and testing the thermochromic ink

itself. The fabric sleeve also frequently detached from the housing. Our solution to both problems was to print the cases with small holes that allowed the ribbon cables and fabric sleeves to be sewn directly to the housing, as shown in the top image. I am not sure how common it is for 3D-printed parts to be sewn to fabric now, but at the time it felt like novel and exciting innovation. The lower image shows testing of the fabric sleeve's thermochromic ink: one section has just been removed from a fridge, while another has been wrapped around a mug of hot coffee.

Image 14.10: Batch-produced parts and devices.

The final step in our process was to batch-produce the Indoor Weather Stations. My memory of this stage has faded significantly over the years, but I do recall everyone on the team signing a written statement, promising to "never, ever build any more Indoor Weather Stations, no matter what", and pinning it to the studio wall. The gesture was half-joking, but with a seriously earnest undertone. I do not feel like this now, and I have since tackled far more elaborate, complicated, and downright difficult projects, so much so that this one seems like a walk in the park in retrospect. However, as can be seen in this image, there were a lot of devices and parts filling our studio. Perhaps it was harder than I remember.

Image 14.10
Batch-produced parts and devices

14.3 Summary

This chapter revisits the Indoor Weather Stations project by reflecting on *Attention to Detail*, the photo essay that first documented its design process over a decade ago. What was once a novel provocation can now be read as a historical artefact, one that serves as a reminder of the depth of detail involved in bringing the Indoor Weather Stations to life, and of the significance of working at scale for the first time. Much like a Record Store Day B-side compilation, my aim here has been to extend that work by offering a glimpse into the original thinking behind producing visual contributions for an HCI audience, alongside a more candid, behind-the-scenes account of how the project was actually made.

Beyond revisiting this particular project, I hope this chapter also reinforces the continued importance of documenting the design process. Many of the nuanced layers of work that shape design research remain largely unseen, yet are integral to the discipline. Written accounts can easily smooth over these details. Writing about over-complicated electronics, the perils of delamination, or sewing fabric to 3D-printed components often fails to convey the lived experience involved, and can flatten those moments into a list of technicalities. At the time, we felt that visual documentation could capture these aspects more effectively. While our initial attempt to experiment with the photo essay format may only just have survived the review process, it is encouraging to see that other designers and readers now appear to agree, given how widely pictorials are used today.

Acknowledgments The research project presented here were supported European Research Council, grant agreement ID: 226528. I wish to offer special thanks to the co-authors of *Attention to Detail*, Nadine Jarvis and David Cameron.

References

Andy Boucher. 2023. Research Products at Scale: Learnings from Designing Devices in Multiples of Ones, Tens, Hundreds and Thousands. In *Proceedings of the 2023 CHI Conference on Human Factors in Computing Systems* (CHI '23). Association for Computing Machinery, New York, NY, USA, Article 177, 1–15. https://doi.org/10.1145/3544548.3581540

William Gaver, John Bowers, Tobie Kerridge, Andy Boucher, and Nadine Jarvis. 2009. Anatomy of a Failure: How we Knew When our Design Went wrong, and What We Learned From It. In *Proceedings of the SIGCHI Conference on Human Factors in Computing Systems* (CHI '09). Association for Computing Machinery, New York, NY, USA, 2213–2222. https://doi.org/10.1145/1518701.1519040.

William W. Gaver, John Bowers, Kirsten Boehner, Andy Boucher, David W.T. Cameron, Mark Hauenstein, Nadine Jarvis, and Sarah Pennington. 2013. Indoor Weather Stations: Investigating a Ludic Approach to Environmental HCI Through Batch Prototyping. In *Proceedings of the SIGCHI Conference on Human Factors in Computing Systems* (CHI '13). Association for Computing Machinery, New York, NY, USA, 3451–3460. https://doi.org/10.1145/2470654.2466474

Nadine Jarvis, David Cameron, and Andy Boucher. 2012. Attention to Detail: Annotations of a Design Process. In *Proceedings of the 7th Nordic Conference on Human-Computer Interaction: Making Sense Through Design* (NordiCHI '12). Association for Computing Machinery, New York, NY, USA, 11–20. *https://doi.org/10.1145/2399016.2399019*

Nicolas Villar, James Scott, and Steve Hodges. 2010. Prototyping with Microsoft.net Gadgeteer. In *Proceedings of the fifth international conference on Tangible, embedded, and embodied interaction* (TEI '11). Association for Computing Machinery, New York, NY, USA, 377–380. https://doi.org/10.1145/1935701.1935790

Part V
For Educators

Chapter 15
For Educators: Design Challenge Based Learning, Quilts, Promptosketches and Promptotypes

Shunying An Blevis and Eli Blevis

Abstract In this chapter, we describe our practice of design challenge-based learning and other constructivist classroom approaches over more than a decade of teaching. We include several examples of projects done as assigned in our classes that relate to *seeing* interaction design, in the hope they are useful examples for others wishing to include visual practices in their curricula.

15.1 Contributions

We present several examples of work by participants in our classes as particularly interesting examples of deliverables for visually oriented projects in interaction design classes. Two of these examples have appeared already in ACM interactions.

15.2 Design Challenge Based Learning

According to Bada (2015), constructivist learning theory holds that learners learn best when they construct knowledge rather than just passively taking in information. Students construct knowledge or learn "by fitting new information together with what they already know." Considering "the learner as an active agent in the process of knowledge acquisition," the constructivist view of learning often recommends

---

With work by Shih-Ting Lin, Jason Lu, Kelly Lucksanawichien, Qingran Ni, Szu-Chin Wang, and Cassiopeia Winslow

---

S. A. Blevis · E. Blevis (✉)
Luddy School of Informatics, Computing, and Engineering, Indiana University, Bloomington, IN, USA
e-mail: eblevis@iu.edu

S. A. Blevis
e-mail: san@iu.edu

S. An Blevis et al. (eds.), *Seeing Interaction Design*, Design Research Foundations,
https://doi.org/10.1007/978-3-032-14200-9_15

using hands-on experience, reflection, and reciprocal teaching and learning to foster active learning.

The chapter shares materials that may be used for teaching how to see interaction design through constructivist classroom activities and assignments: from (a) how to reflect on or relate concepts with diagrams, to (b) how to incorporate others' perspectives with digital quilting, to (c) how to visually represent new knowledge and meaning, to (d) how to visualize new concepts using various image production techniques, to (e) how to use Generative AI to aid design process—that we term herein as *Promptotyping*.

We use the term *design challenge-based learning* (DCBL) to name our particular practice of constructivist learning. In Blevis (2010), DCBL is described as follows:

> The core idea of DCBL is to present designers with humanity- and life-centered issues-based design research and design-concept challenges in the arena of HCI and design that are manifest as (i) individual, collaborative, and competitive activities involving public presentation and critique; (ii) implicit rather than explicit inclusion of rigorous concepts in the service of motivated design-challenge goals; and (iii) linked pairs of research and concept projects that prompt students to practice ensuring that their concepts follow from research insights and that their research insights lead to concepts.

For additional descriptions of DCBL, see also Blevis (2012). For additional descriptions of constructivism in learning, (see also Novak, 1990; Askell-Williams et al., 2012) for examples. There is a broad literature on constructivist learning in education.

Note that we use the term *participants*, rather than *students* or *learners*. We also use the term *tile* interchangeably with *slide*.

Image 15.1 Self Portraits, Zhouxing (Jason) Lu, Photographer, Eli Blevis, Curator/Editor, Photographs, 2015

As a first project in many of our classes, we ask participants to present a visual self-portrait. This is one form of a design challenge that relates to Seeing Interaction Design. Participants present their self-portraits in small groups or to the entire assembly, depending on the class size. This works not only as a means of constructivist learning, but also as a means of having participants introduce themselves one to another and to us.

In 2015, Zhouxing Lu create the triptych of self-portraits appearing in this image. In the top image, Lu investigates the self who did not know about the non-smoking rules on our Bloomington, Indiana campus. His two hands hold up a picture frame as if to cover his face. The frame is actually transparent, revealing a No Smoking sign on a limestone wall behind him. His face is also missing and part of the transparent reveal. This surrealist image shows Lu's vanishing sense of self in the face of cultural context and misunderstanding.

In the middle image, Lu is gathered around a table with a project team for another class which taught hand signal protocols owing to McCarthy and McCarthy (n.d.) for efficiently developing consensus within a team. In the image, everyone on the team is giving a thumbs up with outstretched arms. Lu himself has a third arm emanating from his abdomen through the same empty picture frame in the top image. His third outstretched arm offers a thumbs down gesture, denoting that he actually disagrees with the team, but is afraid to indicate so, regardless of the spirit of the protocols.

In the bottom image, Lu is confidently drawing on a whiteboard for an unseen audience in a classroom. Now he is holding the transparency inducing picture frame at his waist, revealing another version of himself as a very shy person cowering in the corner of the room.

All three pictures are very well-conceived and produced. They were published in (Lu, 2015), as part of the Visual Thinking Gallery edited/curated by author Eli Blevis. *Publishable* is the gold standard for DCBL project work.

Image 15.2 Hired, Cassiopeia Winslow, Photographer, Eli Blevis, Curator/Editor, Photograph, 2014

Another project we ask participants to undertake concerns picturing inclusivity. In *Hired*, Cassiopeia Winslow cleverly shows a pregnancy test stick indicating that someone may be hired, because she is not pregnant according to the test. The legend on the test stick shows the two possible states of outcome for the test, hired, or not hired. This picture is also very well-conceived and produced. It also appears in the Visual Thinking Gallery as (Winslow, 2014).

Image 15.3 Digital Quilt: Reflection on a talk by Jeff Bardzell, Qingran Ni and Jiawei Wang, Artwork and Photograph, 2020

In an introductory class, we asked participants to provide a single tile (or slide) summarizing their reflections on the talk of a guest speaker. In this case, the guest speaker was Jeff Bardzell speaking on the subject of what he believes it means to be a designer in HCI. His talk covered notions such as meaningful technology, design rationality, usability, user experience (UX), cognitive models, accountability, design thinking, framing, and solving, information processing, ethnography, and Fitt's Law.

The individual tiles by each participant were collected and presented in class. A single tile was chosen as a key image—in this case the image by Qingran Ni who uses an image of herself by Jiawei Wang as part of her collage approach to the project. We use a program (AndreaMosaic) to generate a digital quilt composed of all or nearly all of the tiles submitted by the participants.

This image shows the resulting Mosaic (or Quilt) based on Qingran Ni's tile.

Image 15.4 Tile: Reflection on a talk by Jeff Bardzell, Qingran Ni and Jiawei Wang, Digital Artwork and Photograph, 2020

This image shows Qingran Ni's original tile. Against a landscape orientation mustard colored background, she places a portrait mode collage. The background elements of her collage are irregular shapes in an assortment of colors with the notions from the talk delimited above as titles distributed over the shapes. Above these elements is a cutout image of herself striking a vogue-like pose. Above this in script is the word "Designer?" with a question mark. At the base of the portrait mode collage is the text "What does it take and how to become a real US designer?".

Image 15.5 Tile: Reflection on a talk by Jeff Bardzell, Szu-Chin Wang, Digital Artwork, 2020

One of the tiles that serves as a constituent of the quilt in Image 15.3 is the work of Szu-Chin Wang. Wang's reflection on the talk by Jeff Bardzell takes a different, more gestalt like approach. She visualizes her understanding of the talk as a graphic flower with bright pastel colors growing on a cream-colored background. A blue watering-can hovers above and to the right of the flower. Water droplets appear in blue shapes. The acronym HCId (for Human–Computer Interaction Design, the name of our program) appears in four different type styles in the bottom right corner of the tile. Hovering over the tiles to find Szu-Chin Wang's tile may be entertaining. Clicking on the tile once found yields a new tab with Wang's original contribution.

Image 15.6 Tile: Reflection on a talk by Jeff Bardzell, Shih-Ting Lin, Digital Artwork, 2020

In this image, we share another tile by participant Shih-Ting Lin which appears as part of the mosaic quilt of Image 15.3. Lin contributes an iceberg diagram complete with a diver in blue tones. Lin's account of Bardzell's talk lists collaboration, evaluation methods, prototyping, wireframing, and UX research above the water line. Her account lists terms conflict management, critical thinking, design argument, leadership, diversity sensitive, and the phrase "and more …" That is, Lin's diagrammatic tile lists basic interaction design skills above the water line, and more abstract notions below.

Image 15.7 Self-Portrait and E-Waste, Kelly Lucksanawichien, Generative AI Sketch and Prototype, 2023

Eight years after Zhouxing Lu's self-portrait described above, Kelly Lucksanawichien contributes an AI generated self-portrait sketch for a similar assignment. Below this, she contributes an AI-generated 3d representation of a modular laptop. The prompts for these tiles are "A line drawing of a young lady cooking in front of a stovetop" and "A mockup of a modular laptop that could help reduce e-waste by allowing users to replace needed parts." The former may be termed a *promptosketch.* The latter may be terms a *promptotype.*

15.3 Summary

Visual projects may be assigned in interaction design classes using the paradigm we describe here as design challenge-based learning (DCBL) or other paradigms for learning. The example projects we include here include self-portraits, inclusivity, reflections on a guest speaker's talk, as well as *promptosketching* and *promptoyping* using generative AI.

Image 15.1 Self portraits, Zhouxing (Jason) Lu, photographer, Eli Blevis, curator/editor, photographs, 2015

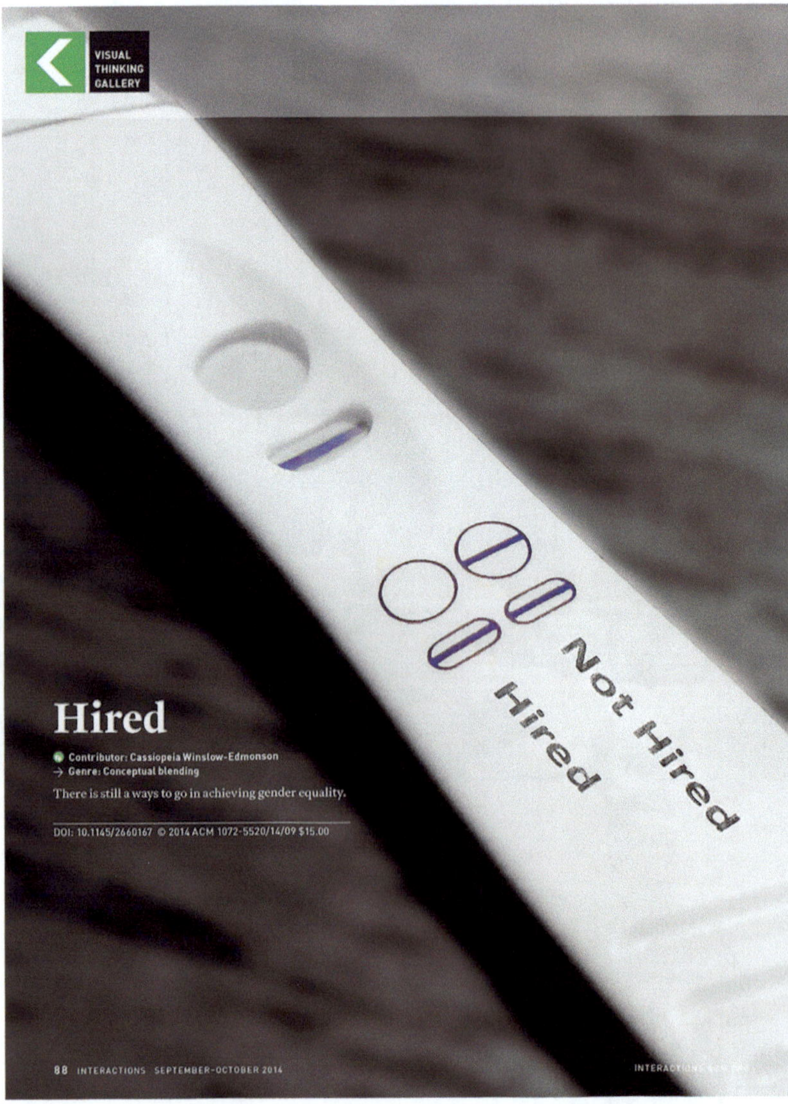

Image 15.2 Hired, Cassiopeia Winslow, photographer, Eli Blevis, curator/editor, photograph, 2014

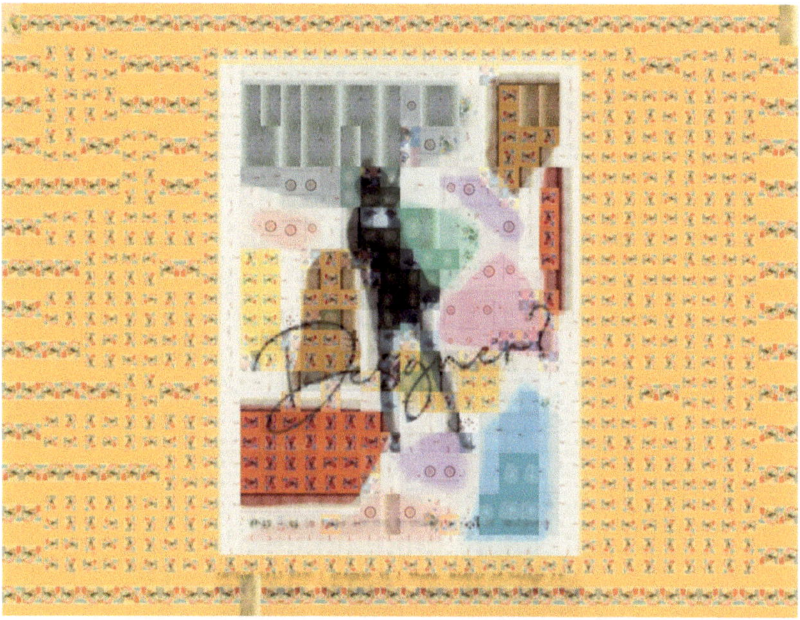

Image 15.3 Digital quilt: Reflection on a talk by Jeff Bardzell, Qingran Ni and Jiawei Wang, artwork and photograph, 2020

Image 15.4 Tile: Reflection on a talk by Jeff Bardzell, Qingran Ni and Jiawei Wang, digital artwork and photograph, 2020

Image 15.5 Tile: Reflection on a talk by Jeff Bardzell, Szu-Chin Wang, digital artwork, 2020

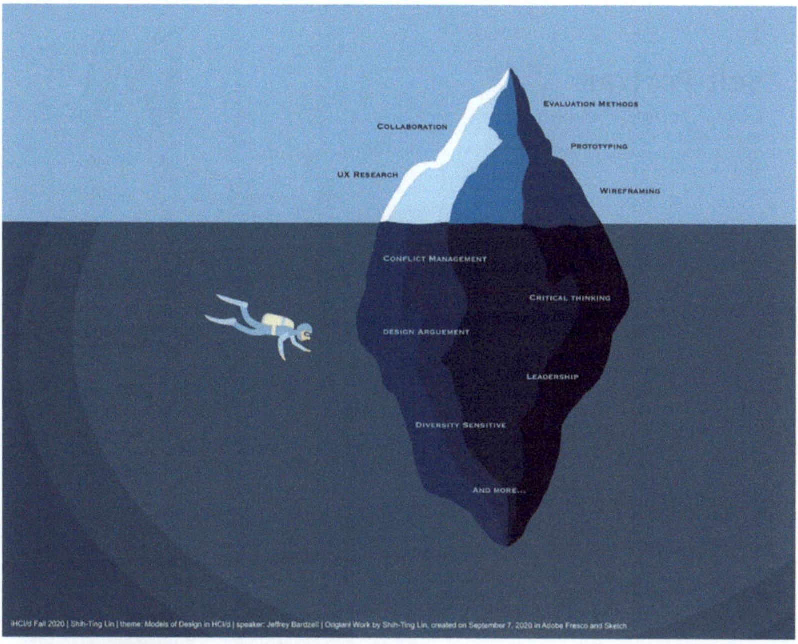

Image 15.6 Tile: Reflection on a talk by Jeff Bardzell, Shih-Ting Lin, digital artwork, 2020

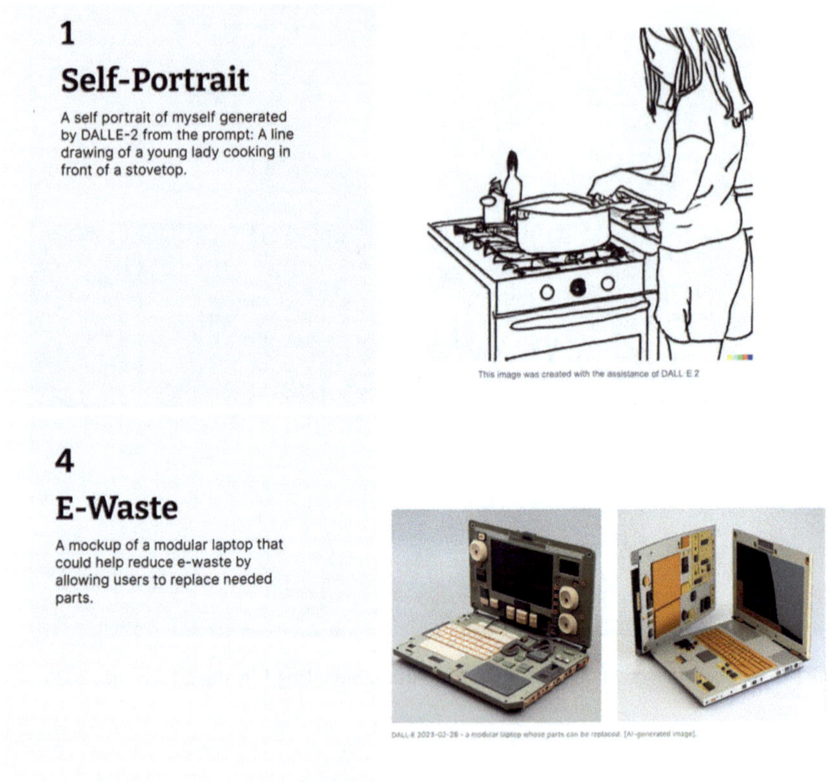

Image 15.7 Self-portrait and e-waste, Kelly Lucksanawichien, generative AI sketch and prototype, 2023

Acknowledgements Many thanks to those who have given us permission to reference and reproduce their work here.

References

Askell-Williams, H., Lawson, M. J., & Skrzypiec, G. (2012). Scaffolding Cognitive and Metacognitive Strategy Instruction in Regular Class Lessons. *Instructional Science, 40*(2012), 413–443.

Bada, S. O. (2015). Constructivism Learning Theory: A Paradigm for Teaching and Learning. *Journal of Research & Method in Education, 5*(6), 66–70.

Blevis, E. (2010, May). Design Challenge-Based Learning (DCBL) and Sustainable Pedagogical Practice. *Interactions, 17*(3), 64–69.

Blevis, E. (2012). The One Room School House & Design Challenge Based Learning for Design Oriented HCI Education: Initial Results, Reflective Hypotheses, & Collaborative Issues. In *2012 International Conference on Collaboration Technologies and Systems (CTS)* (pp. 359–366). IEEE.

Lu, Z. (2015, May—June). Self-portraits. *Interactions, 22*(3), 80. https://doi.org/10.1145/2754307
McCarthy, J., & McCarthy, M. (n.d.). *The Core Protocols V.3.03.* McCarthy Technologies, Inc. Retrieved April 01, 2024 from www.mccarthyshow.com
Novak, J. D. (1990). Concept Maps and Vee Diagrams: Two Metacognitive Tools to Facilitate Meaningful Learning. *Instructional Science, 19*(1990), 29–52.
Winslow, C. (2014, September–October). Hired. *Interactions, 21*(5), 88. https://doi.org/10.1145/2660167

[illegible]

Chapter 16
Examining AI-Generated Art in the Educational Context

Akesha Horton

Abstract This chapter provides a brief survey of the academic literature surrounding some of the essential questions related to AI generated art for art and craft educators. The author also examines practical advantages and challenges of using generative AI tools for design as an educator as well as a professional crafter.

16.1 Introduction

Art has traditionally been considered an artifact of human civilization. (Morriss-Kay, 2010) describes art as "…a defining characteristics of the human species" (p. 158). Visual artists have used computers in their practice since 1950 (Hencz, 2022). While artificial intelligence (AI) tools have been used to create AI art since 1973 (Zheng et al., 2022) recent advancements in artificial intelligence have broadened access to the creation of Intelligent Art (IA) or art created with artificial intelligence, to a wide array of people. It has also expanded the possibilities of what can be created. However, (Pandey et al., 2022) reminds us that "Deep Learning (DL) technology is still not widely used in Media art and crafts education", especially in the case of traditional complex art forms. The rise in intelligent art created with generative AI (GAI) has opened the door to new conversations that have raised more questions than answers in arts communities as well as among art educators. GAI is defined as "artificial intelligence that can generate novel content, rather than simply analyzing or acting on existing data like expert systems" (Gozalo-Brizuela & Garrido-Merchan, 2023). As GAI allows the average person with no training in art to instantly create art, crafts, and models with text prompts with the fidelity of an expert. Educators have become increasingly interested in how IA redefines the discipline. This offers artists and educators the opportunity to determine the 'essential questions' that should be considered when integrating AI into the praxis of art and art education.

A. Horton (✉)
Indiana University, Bloomington, Indiana, US
e-mail: akmhorto@iu.edu

S. An Blevis et al. (eds.), *Seeing Interaction Design*, Design Research Foundations,
https://doi.org/10.1007/978-3-032-14200-9_16

Essential questions as defined by (McTighe & Wiggins, 2013) are questions that encourage deeper inquiry about a discipline. They are intended to initiate a conversation about a topic as opposed to providing a direct answer. A few of the reasons McTighe & Wiggins encourage the use of essential questions is because they prompt intellectual engagement within a discipline. Through such engagement, educators can develop standards and priorities when teaching and identify opportunities for intra- and interdisciplinary connections. A review of the literature identifies a few essential questions for artists and art educators to contemplate. (Oppenlaender, 2022) asks, "If anybody can produce digital images that resemble masterful pieces of art simply by feeding textual prompts into a "black box" system, is text-based generative art creative? What is the nature of the human creativity involved in generating images with text-based generative systems?" (Mazzone & Elgammal, 2019) ponder, 'If artificial intelligence is used to create images, can the final product really be thought of as art? Should there be a threshold of influence over the final product that an artist needs to wield?' ("How AI Is Transforming the Creative Industries," 2021) Or should AI be credited as a collaborator in the creative process? "In the creative industries, such as music, AI is more likely to help artists make new kinds of art, rather than replace their jobs." (The Economist, 2021). The article questions, however, whether AI generative designs will be more imitation or innovation? Does integrating these tools diminish the creativity of those working with it? How will it impact the intellectual property of an item created in collaboration with AI tools? While (Tedre et al., 2023) explore questions about the data sources used to create the neural networks that generate the images, as well as what challenges and limitations exist when using AI in the design process?

There are also important practical (yet non-essential) questions that are helpful to examine as well. (Chiu et al., 2022), asks whether AI tools can be used to successfully train crafters on art and craft practices, while (Wenjing & Cai, 2023) examine which attributes are important for AI generative technology to have to be useful to artists. While the courts have ruled that only humans can hold a copyright (Guy, 2023), are there innovative approaches to creating IA that may challenge this stance (*Is This Art?*, 2018)?

16.2 Contributions

In this chapter I position myself as an artist, educator, and student. I create in the medium of paper. I practice paper quilling. Paper Quilling is the art of creating 2-D and 3-D designs out of thin paper spirals and curls (Ceceri, 2015). I have been an active member of two art guilds for about five years: The Guild of American Papercutters and the North American Quilling Guild. In 2019, I was presented with the idea of using computer software to create 3-D quilling models before committing them to paper using a 3-D software package called Houdini (Entagma, 2019).

While I have used 3-D software (Maya, Unity) to create animations, using it for quilling seemed inefficient. It would take more time to create the 3-D model than it

Image 16.1 Image of Sydney opera house on a spring morning generated by Midjourney. The prompt: Sydney opera house at morning on a spring day

would to create the paper quilled art piece. Fast forward a just a couple of years to Deep Dream Generator Midjourney, and the world of command line art prompts. The ability to efficiently create models was cut down to seconds, but essential questions arose related to how IA could be used in the creative process. I share below some of my designs created in Midjourney, as well as some of the essential (and non-essential questions), that arose during the process (Images 16.1–16.8).

16.3 Summary

This chapter provides a brief summary of the essential and non-essential questions that an artist and art educator should consider when using generative AI in their creative process. These questions are intended to start conversations within the discipline to determine the best way to use generative AI tools for creative and educative

Image 16.2 Images of Sydney opera house on a spring morning generated in the style of quilling by Midjourney. In theory Midjourney should have used the picture generated in Image 16.1 to create the featured compilation. However, Midjourney has interpreted its own creation (Image 16.1) as a flower as opposed to the Sydney opera house when asked to recreate the image in a different art style. Midjourney's struggle to create a quilled version of the Sydney Opera House happens because of a tug-of-war on its internal mathematical map, known as a latent space. This map is where the AI groups similar ideas together. In this case, the smooth, giant curves of the building's architecture are fighting against the tiny, busy coils of the quilling style (Patel, 2025; Schaerf, 2024). The problem gets worse because of semantic dilution. When you add extra words like paper art or kirigami, the AI gets confused and blends the styles together. The white, overlapping sails of the Opera House share a nearly identical geometry with the layered petals of a flower. In the AI's training data, the quilling style is also much more strongly associated with flowers than with buildings. When asked to combine them, the model prioritizes the quilling pattern and slides the image toward a floral cluster because those shapes fit the artistic style more naturally than a rigid building (Patel, 2025). Finally, the AI fails because of a denoising conflict. During the process of turning static into a clear picture, the model has a hard time focusing on two things at once: the big shape of the building and the millions of tiny paper loops. To make it work, the AI summarizes the image by choosing simpler shapes, which is why the fine details of the quilling often disappear (Wang et al., 2025)

Image 16.3 Images of Sydney opera house on a Spring morning generated in the style of quilling by Midjourney. The prompt: Sydney opera house at sunset on a spring day in the style of quilling. (Similar results with the prompts of 'quilled paper art' and 'paper kirigami quilling'). Only one of the images generated show many of the iconic features of the Sydney opera house. Also, while the image has characteristics of paper art, it has few attributes of traditional paper quilling. Is the lack of quilling characteristics due to the dataset? Midjourney's failure to accurately render traditional quilling in the Sydney Opera House prompt likely stems from a tug-of-war within its latent space, defined as the high-dimensional mathematical environment where the model maps relationships between abstract concepts (Patel, 2025; Schaerf, 2024). In this instance, the smooth, architectural geometry of the landmark competes with the high-frequency, intricate coils characteristic of quilling. This conflict is further exacerbated by semantic dilution, which is a phenomenon where including broader terms like paper art or kirigami causes textual embeddings to blend, steering the model toward a generic 3D papercraft aesthetic rather than specific filigree. Ultimately, the model defaults to summarizing the texture into simpler shapes due to a denoising conflict; the resolution required to maintain both architectural recognition and dense, microscopic paper coils creates a complex data-association challenge that the generative backward process struggles to reconcile (Wang et al., 2025)

Image 16.4 Image of Sydney opera house in the style of quilling by Midjourney. The prompt: Sydney opera house in the style of quilling. My goal was to simplify the prompt to focus on the Opera House. However, the flowers (which were not part of the prompt) were more prominent than the building in most of the images produced

purposes. While tools like Midjourney and Deep Dream can create complete pieces of art in seconds, this does not mean the results will automatically yield the vision the artist has in mind. This chapter summarizes the essential questions artists and art educators should consider when using generative AI. These questions aim to guide the discipline in determining the best use of these tools for creative and educational purposes. While tools like Midjourney and Adobe Firefly generate images quickly, they often fail to match an artist's specific vision due to technical limitations. The struggle to render the Sydney Opera House as paper quilling illustrates these hurdles. Models often face a denoising conflict where they cannot reconcile intricate textures with complex architecture. This leads to failures like semantic displacement, where the AI prioritizes common patterns like flowers over the requested subject. These results show that prompting is more than just choosing words. It is a technical negotiation with the model's latent space. Achieving a specific goal requires time, research,

Image 16.5 Images of Sydney opera house on a Spring morning generated in the style of papercut by Midjourney. The prompt: https://s.mj.run/DkUrslavdNc in the style of papercut was used for the image on the left. The link is the picture featured in Image 16.1. The prompt: Sydney opera house at sunset on a spring day in the style of papercut. This time the images match the style of a papercut image and can be used as models for creating a layered image using text weight or card stock paper

Image 16.6 Images of Sydney opera house on a Spring morning generated in the style of quilling by Adobe Firefly. The prompt: Paper quilling of Sydney opera house in Sydney Australia was used for the image on the left. The prompt: Sydney opera house at sunset on a spring day in the style of quilling was used for the picture on the right. The image on the left shows the skyline of a city in the distance with paper quilling strips laying in the grass. The image on the right shows s picture of traditionally quilled flowers and shapes. The center features what appears to be the sun above a grassy field and mountains. This time the generative tool recognizes what traditional paper quilling is, but does not appear to be able to create an image of the Sydney opera. Adobe Firefly's output highlights a spatial grounding failure where the model recognizes the physical properties of quilling but cannot synthesize them into architectural form. This stems from a denoising conflict where the model maintains the integrity of the quilled coils at the expense of the Sydney Opera House's global structure (Wang et al., 2025). Unlike Midjourney's stylistic blending, Firefly experiences semantic displacement. Its latent space contains such strong associations between quilling and botanical subjects that the architectural prompt is overridden by more statistically dominant craft imagery (Patel, 2025; Schaerf, 2024). Consequently, the model defaults to a compositional juxtaposition of the two concepts. It places literal paper strips in the foreground of a city scene rather than executing a unified stylistic transformation of the building itself

Image 16.7 Images of Sydney opera house on water by *DALL·E*. The prompt: Paper quilling of Sydney opera house at sunset on a spring day in the style of quilling. *DALL·E* created a paper model of the Sydney opera house with traditional quilling shapes hovering above the opera house. The background appears to be a photographic image that places the model of the opera house in water. This interpretation provides a very literal depiction of the prompt. All the items from the prompt are represented, but the concept of paper quilling is separated from the image of the opera house

Image 16.8 Image of Sydney opera house. The prompt: Paper quilling of Sydney opera house at sunset on a spring day in the style of quilling. Deep Dream created a paper model of the Sydney opera house comprised of a mixture of traditional quilling shapes and paper cuts. The background appears to show the model of the opera house sitting on a blue base with a greyish blue background. The sunset and spring were not generated in this picture

and an understanding of how AI associates data. Ultimately, an artist may still need to use additional tools to manually correct these structural errors and satisfy their original intent.

References

Ceceri, K. (2015). *Paper Inventions: Machines That Move, Drawings That Light Up, and Wearables and Structures You Can Cut, Fold, and Roll* (1st ed.). Maker Media.

Chiu, M.-C., Hwang, G.-J., Hsia, L.-H., & Shyu, F.-M. (2022). Artificial intelligence-supported art education: A deep learning-based system for promoting university students' artwork appreciation and painting outcomes. *Interactive Learning Environments*, 1–19. https://doi.org/10.1080/10494820.2022.2100426

Entagma (Director). (2019, December 26). *Procedural Modeling In Houdini: Quilling*. https://www.youtube.com/watch?v=v732_IOCvRY

Gozalo-Brizuela, R., & Garrido-Merchan, E. C. (2023). *ChatGPT is Not All You Need. A State of the Art Review of Large Generative AI Models*. https://doi.org/10.48550/ARXIV.2301.04655

Guy, Z. (2023, August 21). *AI-Generated Art Is Not Copyrightable, Judge Rules*. Vulture. https://www.vulture.com/2023/08/ai-art-copyright-ineligible.html

Hencz, A. (2022, April 8). AI Art and How Machines Have Expanded Human Creativity. *Artland Magazine*. https://magazine.artland.com/ai-art/

How AI is transforming the creative industries. (n.d.). *The Economist*. Retrieved July 2, 2023, from https://www.economist.com/films/2021/04/07/how-ai-is-transforming-the-creative-industries

Is this art? AI-generated portrait fetches over $400,000 at auction. (2018, October 29). *New Atlas*. https://newatlas.com/ai-art-auction-obvious-belamy/56984/

Mazzone, M., & Elgammal, A. (2019). Art, Creativity, and the Potential of Artificial Intelligence. *Arts, 8*(1), 26. https://doi.org/10.3390/arts8010026

McTighe, J., & Wiggins, G. P. (2013). *Essential Questions: Opening Doors to Student Understanding*. ASCD.

Morriss-Kay, G. M. (2010). The evolution of human artistic creativity. *Journal of Anatomy, 216*(2), 158–176. https://doi.org/10.1111/j.1469-7580.2009.01160.x

Oppenlaender, J. (2022). *The Creativity of Text-to-Image Generation*. https://doi.org/10.48550/ARXIV.2206.02904

Pandey, A., Janga, V., Aweke Wako, D., Senbato Genale, A., Sundaram, B. B., & Karthika, P. (2022). Deep Learning Technology using New Media on Traditional Arts and Crafts of Complex Forms. *2022 6th International Conference on Trends in Electronics and Informatics (ICOEI)*, 1359–1365. https://doi.org/10.1109/ICOEI53556.2022.9777160

Patel, S. (2025). *The Latent Space Hypothesis: Toward Universal Medical Representation Learning*. arXiv. https://doi.org/10.48550/arxiv.2506.04515

Schaerf, L. (2024). *Reflections on Disentanglement and the Latent Space*. arXiv. https://doi.org/10.48550/arxiv.2410.09094

Tedre, M., Kahila, J., & Vartiainen, H. (2023). *Exploration on How Co-designing with AI Facilitates Critical Evaluation of Ethics of AI in Craft Education*, 2289–2296. https://www.learntechlib.org/primary/p/222124/

The Economist (Director). (2021, April 7). *How AI is Transforming the Creative Industries*. https://www.youtube.com/watch?v=cgYpMYMhzXI

Wang, Y., et al. (2025). Artificial intelligence using a latent diffusion model enables the generation of diverse and potent antimicrobial peptides. *Science Advances, 11*(1). https://doi.org/10.1126/sciadv.adp7171

Wenjing, X., & Cai, Z. (2023). Assessing the Best Art Design Based on Artificial Intelligence and Machine Learning Using GTMA. *Soft Computing, 27*(1), 149–156. https://doi.org/10.1007/s00500-022-07555-1

Zheng, X., Bassir, D., Yang, Y., & Zhou, Z. (2022). Intelligent Art: The Fusion Growth of Artificial Intelligence in Art and Design. *International Journal for Simulation and Multidisciplinary Design Optimization, 13*, 24. https://doi.org/10.1051/smdo/2022015

Chapter 17
Wacky Encodings for Humanities Visualization

Richard Brath, Johanna Drucker, Johannes Liem, Christophe Schuwey, and Joris J. van Zundert

Abstract This research is guided by several convictions. The first is that visualization conventions predispose researchers to see their data in terms of standard types (bar charts, plots, network diagrams, and so on). The second is that getting beyond these conventions through graphical exploration offers a way to rethink data structures as well as analyses. Third, graphic experiments can provide a primary mode of intellectual work for production and analysis of research across fields and disciplines. To explore these assertions, the authors conducted a workshop based on a provocation to design alternatives to conventional visualizations. Participants were asked to create graphic tools that could produce unexpected outcomes. The result was a surprising number of novel innovations, described below, accompanied by a discussion of possible implications for future research either implementing the tools in prototype or adopting their methods for data design.

R. Brath (✉)
Uncharted Software, Toronto, ON, Canada
e-mail: rbrath@unchartedsoftware.com

J. Drucker
Information Studies, University of California-Los Angeles, Los Angeles, CA, USA
e-mail: drucker@gseis.ucla.edu

J. Liem
Center for Cultures and Technologies of Collecting, University for Continuing Education Krems, Krems an Der Donau, Austria
e-mail: johannes.liem@digitalcartography.org

C. Schuwey
Ingénierie du Document, Université de Bretagne Sud, Tressignaux, France
e-mail: christophe.schuwey@univ-ubs.fr

J. J. van Zundert
Computational Literary Studies, Huygens Instituut, Amsterdam, Netherlands
e-mail: joris.van.zundert@huygens.knaw.nl

S. An Blevis et al. (eds.), *Seeing Interaction Design*, Design Research Foundations,
https://doi.org/10.1007/978-3-032-14200-9_17

17.1 Introduction

The design of data visualizations requires transforming data into graphics that follow several well-established conventions. In the usual lifecycle of data to display, this process begins by modeling data so that it will conform to a visualization that has specific graphic attributes that serve equally specific analytic approaches. Common data analyses include comparison of values, percentages of part to whole, change over time, and relationships of kinship, derivation, or affinity. These models anticipate their display modes in geometric shapes (bar charts, pie charts, bubble diagrams, tree maps and other formats), lines (continuous curves), plots (points in a grid) networks (node and spoke), and other forms (whisker diagrams etc.) that express information according to a shape's size or color mapped to data values.

All of these formalized designs for data visualizations, as well as their encoding in computer software, have their origins in quantitative data analysis and do not suit needs of domains where multiple perspectives and multiple interpretations may be required such as the humanities (see Ware [2000] for a standard reference; and Drucker [2014] for discussion of humanities approaches). Taken one step farther, we can argue that in standard practice, display drives research through a process of reverse engineering so that graphic output actually structures analysis from the beginning. In pedagogical situations, this realization becomes evident very early in the design practice as students often begin their work by imagining they are intending to create a network diagram, bar chart, or other format as their result. In the process, their conception of a project is intellectually constrained by that conception.

But what if the design process could be upended and innovative graphics could expose novel approaches to structuring and modeling data? To break-out of visualization design thinking as constrained by quantitative analysis and computational systems, we explored the possibilities of graphical expression in both production and presentation through a workshop in which participants were invited to produce novel visualization encodings by hand-drawing expressive, non-standard, representations and engage in community reinterpretation via a short workshop.

Our contribution is the demonstration that given an opportunity to think beyond conventional data visualizations, researchers will come up with innovations that could be used to rethink data structures and models through graphical means.

17.2 Background

The current notion of data visualization design space is highly constrained by its formalizations. For example, psychological testing is used to validate which of several visual attributes are perceived more quickly, or estimated more accurately (Ware, 2000); statistics researchers determined which data types are more effective with which visual attributes (Cleveland & McGill, 1985; Heer & Bostock, 2010); computer science has created rules for how the encodings fit together to form visualizations and so on (Wilkinson, 1999). While there are many variations on this atomic, building-block conceptualization of visualization, some common foundations can be traced back to Jacques Bertin's *Semiology of Graphics* (Bertin, 1983). Furthermore, the computational tools with which we create data visualizations, such as Excel, R (a statistical analysis tool) or D3.js (a programming library) are either constrained by these prior definitions, or simply map quantitative values to the most simple attributes of the underlying computer graphics system (e.g. geometric shapes, primary colors and sizes, vs., say, squiggly lines with variable weight and inflection or any of the hundreds of irregular shapes in the natural and visual world).

This data visualization design space does not fit all needs. This highly constrained design space has its origins in quantitative data analysis and may not be appropriate for all kinds of data visualizations and research agendas. Moreover, it constrains conceptualization by predisposing researchers to imagine their outcome in alignment with its forms. Humanities data, such as books, poetry, paintings, sculpture, scrapbooks, and so forth, aren't quantitative. Interpretation is a complex rather than linear process. Just because it may be feasible to extract quantitative data (such as dates, authors, locations) from humanistic documents does not necessarily mean that that's what should be visualized. Word clouds miss the opportunity to convey much information (Brath, 2021). As per William Bruce Cameron, "Not everything that can be counted counts." (Cameron, 1963)

Singular interpretation of the data does not hold in domains such as humanities. Interpretation in the humanities depends on the viewer's perspective. Time, place, ethnicity, background are different lenses: the interpretation is different for an indigenous person, a long-haul truck driver, a back-country skier, or a new immigrant. Quantitative data visualization is not designed to address multiple simultaneous hypotheses and interpretation or to inflect the graphics with a point of view. In addition, the use of visualizations as a display, rather than as a primary mode of intellectual inquiry, reduces the graphics to a secondary role rather than making use of their capacity to produce knowledge directly.

Humanistic interpretive activity does not necessarily fit well with traditional computational quantitative visualization in a linear method sequencing hypothesis, data, development, and evaluation. Furthermore, current visualization does not work well with heterogeneous data nor polyvalent interpretation. At the highest level, this group felt that the issue could be best addressed through a prompt for speculative case studies, and reference to case studies related to humanities visualization but not perceived as such.

Graphical invention: While the humanistic examples offer possibilities for rethinking data models, they do not take their inspiration from graphical innovations. We wanted to push further than thinking about analysis that resists quantitative analysis and push towards graphical innovations that would lead research. The myth of transparency, so regularly endorsed within the data visualization community, conceals the processes of abstraction, reduction, and representation through which visualizations come about. Rather than a constrained system with a formal grammar to be assembled in the best configuration for the given task, we decided to invert the process and explore other possibilities of graphical expression in both production and presentation of information.

With this in mind, we created a provocation to create speculative visualizations, the "wacky encoding exercise." The first step in this process was to provoke a diverse set of novel visual ideas that can be utilized in successive exercises, for example to create speculative case studies that would create non-traditional datasets as well as serve fields that have many inherently non-quantitative methods, such as the humanities.

17.3 Wacky Encoding Exercise

To hack around the constraints of formal visualization grammars, we asked workshop participants to conceptualize a novel transformation palette—i.e. constructive design research (Koskinen et al., 2011). This produced a suite of many different ways to produce information through interactive representations (and vice versa).

Our point-of-departure was the notion of strange and wonderful paint-brushes in drawing tools geared for children, e.g. KidPix: "The computer can make marks in a variety of interesting ways that traditional tools can't. It can also do a good job of creating comic tools. This is where the wacky brushes came from." (Hickman, 2019)

We encouraged our participants to think that any representation could be fair game for breaking conventions. Individuals in our group guessed about smoke, reflections, historic lithographs, but to generate truly innovative perspectives, we use a public workshop to provoke novel responses through participant-created, on-the-fly, hand-drawn sketches.

Image 17.1:

Sketching provocation: We provoked 30 of our peers at a recent workshop on visualization and the digital humanities: *"Remember when paint programs had cool brushes that create wacky effects? Sketch a tool that has unexpected behaviors. Make a visual effect. Then imagine what kind of data it could represent."*

Participants created 32 responses after 20 min of drawing (Image 17.1). As hoped for, a wide variety of responses were generated, such as examples breaking the 2D plane; expressiveness ranging from diagrams to naturalistic to comic; varying amounts of textual annotation from none to paragraphs.

Community interpretation: Individuals took turns selecting one drawing at random, presenting it, and proceeding to have a brief community discussion and interpretation. This was unstructured, all participants could comment.

Image 17.2:

Title: GPT Connectors

Descriptive text: Draw lines between any entities in any shape. GPT starts with some source text from the starting entity and then the generative program adds variants of the text as you draw the connection between the sources.

Community interpretation: This could work like six degrees of Kevin Bacon, extended to any humanities subjects. The connections could be close, say, connecting Shakespeare and Christopher Marlowe. Or distant, between Macbeth and Robert

Image 17.1 Exercise in progress

Image 17.2 GPT connectors

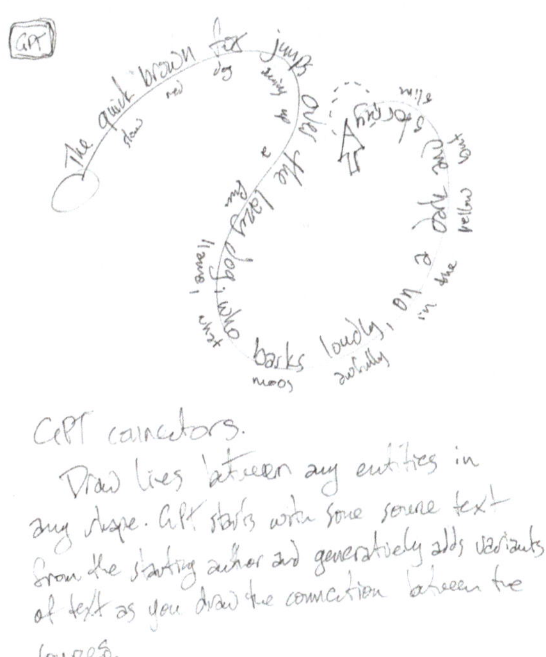

Crumb? Or a raven and a writing desk? The tool raises some interesting questions: How would a GPT connect them, what insight would be gained? Given GPT's are non-deterministic, multiple lines between the same pairs of entities will generate different answers. How would shorter and longer lines pan out? Does a longer line instruct the GPT to generate a longer text, or rather a more exotic variant of the text? What property of the generator might be connected to the squiggliness of a line? The tool seems to inspire the question: what properties of line-art and text are congruent?

Image 17.3:
Title: Surprise Tool
Descriptive text (on back of paper): This tool has two formats, Basic and Pro.

Basic: Forms, visual representations, symbols are transformed into text.
Pro: insert vis and let it be translated into text.

Community interpretation: The tool can "textify" what someone is drawing. Z-translation of their drawing into a box. First reverse visualization tool: transforms what you draw into text. You can pay for the pro version, insert a visualization and it becomes the visualization in a box.

This tool critiques the simple geometric shapes (in today's visualizations) and their transformation to text; whereas the humanities text may be far more complex than what can be expressed in simplistic shapes.

They say a picture is worth a thousand words. But how many pictures can one word evoke? This tool can be interpreted as a play on the finite relation between shapes and their descriptors, between images and their captions. What creativity is inspired when we are kindly forced by this tool to write words by drawing pictures and when we must paint images by using words?

This tool spells it out! By writing bluntly what is actually there, the tool offers a new perspective on a visualization. Does the intended meaning of the graphic hold when transformed as a text? Does the text descriptions foster new interpretations of the data? It also summons senses beyond sight: if the text is read by an automated voice, does something new emerge?

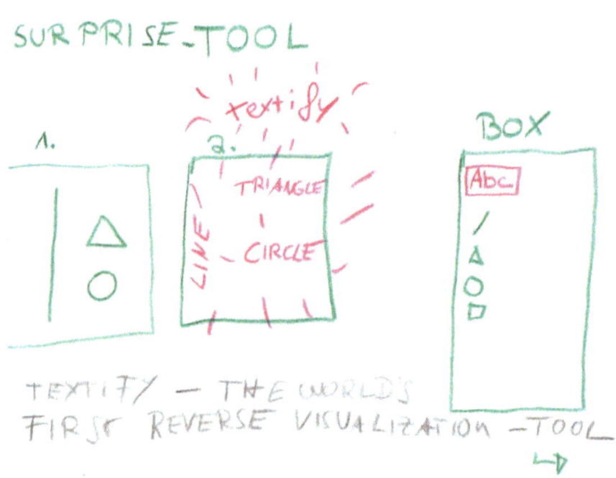

Image 17.3 Surprise Tool—Textify. The back side of this drawing reads: Basic: Forms, visual representations, symbols are transformed into text. Pro: insert vis and let it be translated into text

Image 17.4 Seed ruler

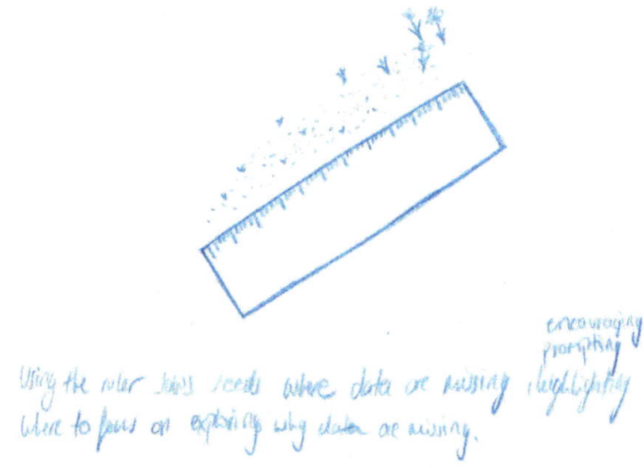

Image 17.4:

Title: N/A

Descriptive text: Using the ruler sows seeds where data are missing, encouraging, prompting, highlighting where to focus on exploring why data are missing.

Community interpretation: It's a ruler that has some detritus along it? Using the ruler sows seeds where data is missing—inviting, exploring, sprouting where data was missing.

Wow! Automated (biologic) techniques for indicating and filling in gaps! There is paper with embedded seeds embedded that will grow if it gets wet. What if you used that kind of paper to print visualizations on? This is related to Dietmar Offenhuber's autographic visualizations which utilize natural, environmental marks (Offenhuber, 2023).

Absence of evidence is not evidence of absence. A rule that is all too often forgotten when researchers interpret large sets of data. Missing data is a pernicious blindspot in all research. What if the analytical tools we use in statistical software packages would paint black swans and unicorns where data was not found? Would we be more inspired to reflect on what might be unseen and missing than when the banal "<NA>" is depicted in a table in Pandas or R?

Image 17.5:

Title: Dagstuhl HCI Semiotic Sensemaking Pipeline

Descriptive text: There's too much academic chatter in this castle. The DHCISSP turns scholarly noise into abstract, expressive, imposing pieces of art. Thanks for your service.

Community interpretation: I like how the pipeline does not stop at the visualization, but continues with a Rorschach test, the textual output of which is converted into an image and assembled into a gallery; reminiscent of Eco's Travels in Hyperreality, wherein the multiple reinterpretations go beyond and may surpass the original. This pipeline—a super popular process model in research software engineering—seems to continue where normal process engineers might have called it a day. Instead of suggesting a final result after a distinct number of precisely defined and scientifical

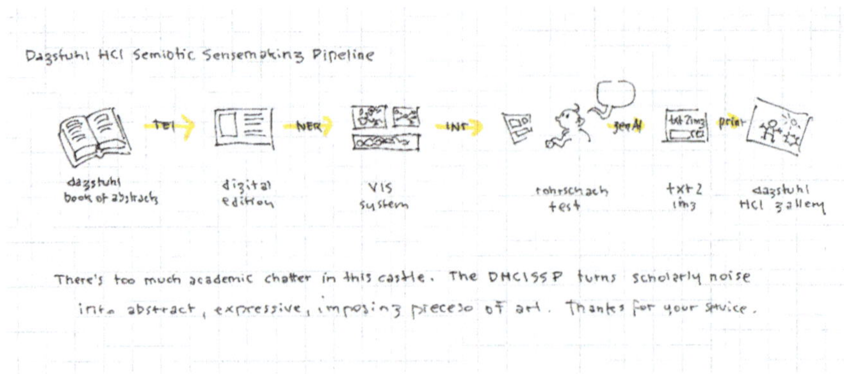

Image 17.5 Semiotic sensemaking pipeline

analytic steps it adds unexpected interpretative operations, ad absurdum? It possibly argues that all operations are interpretative, and it may question the authoritative scientific moves that some transformations in the digital context have been adorned with. How do we decide where the limits of scientifically valid interpretation are? And does formalizing such limits not just hamper creative thinking?

Image 17.6:

Title: Pinball DH

Descriptive text: Point and click to create interactive pinball items—flipper, bouncers, etc. Drop in your source entities—people, places, etc., and they bounce off of the flippers and bumpers to create serendipitous connections between them.

Image 17.6 Pinball DH

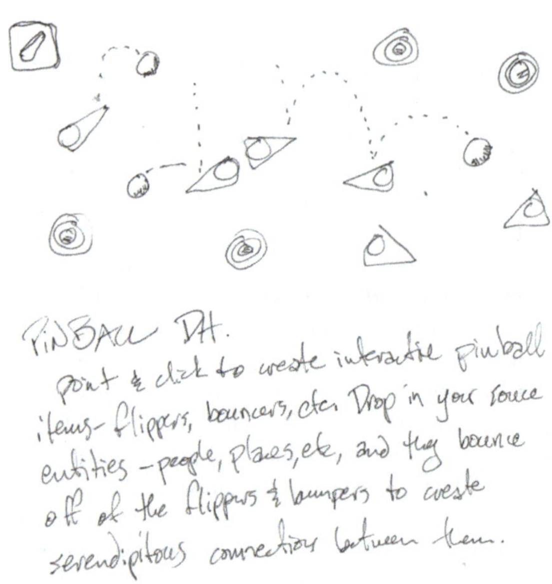

Community interpretation: The kinetic interaction of the pinball machine suggests connections, moments of ever-higher intensity, secondary objectives, and suggests a (gamified) performance of cultural interpretation. Each instantiation is a different interpretation. The pinball gamification of interpretation also invites us to think about the role of chance and arbitrariness that underlies much of cultural analytics interpretation. A mechanism that inspires both good and bad for being on the one hand powerfully creative and on the other possibly dangerously nonsensical. It is also a powerful pedagogical tool to discuss network diagrams limits and meanings, and to experiment how graphical connections do not necessarily mean an actual relationship between items.

Image 17.7:

Title: Funguys

Descriptive text: For every click a specified (=4) number of funguys sprout randomly around a point. Funguys approximate population of other densities.

Community interpretation: The cuteness seems to make it surprisingly engaging. The Funguys inspire images surprisingly congruent to gremlins, minions, goblins, or house elves. They are cute but they also look like they have sufficient agency for potential mischief. Their numbers represent population densities or other distributions. Maybe they represent those properties of populations that mischievously elude the statistician?

As fungus grows on something, Funguys could also point to part that has been abandoned: which areas of data do the Funguys like the most?

Image 17.7 Funguys

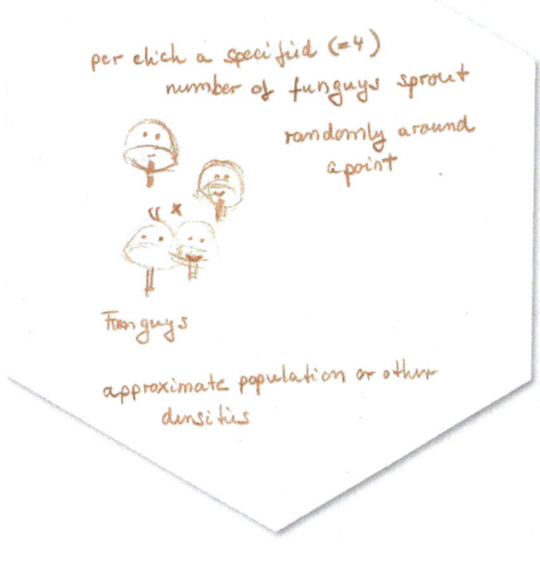

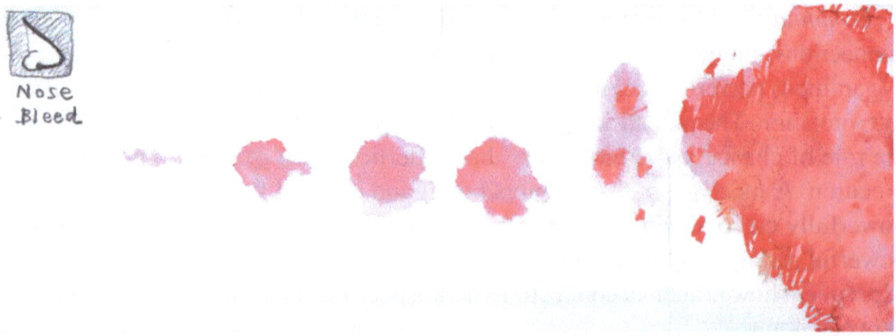

Image 17.8 Nose bleed

Image 17.8:

Title: Nosebleed.

Descriptive text: None needed.

Community interpretation: This literally unfolds, and as you unfold it there's a progressive amount of blood, expanding with circles. This unfolding immediately prompted a strong reaction of disgust. The author commented "It's a 7 year old's technique for making marks: there can be small drops, a pool, or a massacre – which you can apply to show much you hate the work." Which prompted "How much you hate the work," or "How much pain you have put into it." Is this the result if J.G. Ballard made visualizations instead of William Gibson?

It is not too hard to imagine why human–computer interfaces almost never use human biology-based skeuomorphism. This visualization puts any skeumorph into perspective. Remember the tutorial for Mac OS7? (System and 7 Macintosh Basics, www.youtube.com/watch?v=3ScS4OYDfHE) Put it in perspective with the shock of this visualization. It reveals how non-human is a human-like figure (and any kind of skeumorph).

The emotive response for most people is overwhelmingly strong, and, almost fully eclipses the teleological intent of any interface. Body horror evoking Videodrome-like visualizations may have their uses, but they may be hard to determine. Probably use sparingly?

Image 17.9:

Title: Ribbit

Descriptive text: Ribbit enacts the element of surprise. Instead of "flipping" pages, you jump across the text.

Community interpretation. The frog is pensive and enigmatic. It's non-linear access. The was also an interesting reaction by the audience: the room went silent when the frog was "pensive and enigmatic." This seems like a physicalized hyper-link, or like juxtaposing collected fragments for new interpretations in scrapbooks (Moynihan & Putra, 2019).

Most wacky tools seem to be bound together by how they flout the boredom invoked by the all too utilitarian function icons and buttons that we are familiar with

Image 17.9 Ribbit

through the graphical interfaces of our daily dosage of software and apps. Ribbit invites us to break the mold of boringly engaging with data and text.

This is a tool for the twenty-first century attention span. Feeling about to turn to something else? Click on Ribbit.

Image 17.10:

Title: See_thru

Descriptive text: Draw through the drawings of everyone who used this brush before you. There was a strong emotional response from the participants when this was read.

Community interpretation: Like a performance or interpretation on a major work, perception of this visualization will be colored by prior interpretations of the work. With See_thru the work cannot be interpreted only in relation to oneself. This enables the reviewer to see and engage with the work within it's much deeper, broader cultural context, similar to hundreds of visualizations contextualizing in *L'Atlas Molière* (Dealberto et al., 2022).

The profound pondering that the presentation of "See_thru" inspired perhaps reveals how much people seek to connect their creative legacy. Rather than be potentially offended by another intruding on their work, participants seemed to revel about the possibility that their creativity could be reshaped by an unknown future fellow artist.

See_thru summons ghosts and dispels illusions. Against the myth of the solitary genius, the tools sets creativity as being the last dot in a long line: it's impossible to

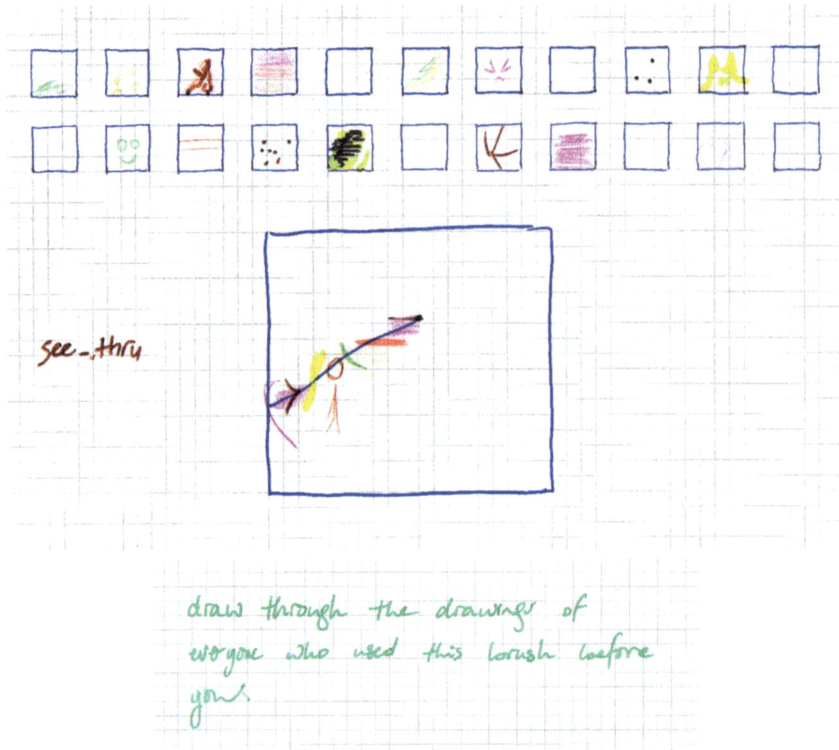

Image 17.10 See_thru

draw a line without summoning everyone who came before and surrounds us. The line exists because of all of them.

Image 17.11:

Title: The Ant Pen

Descriptive text (back of paper): 1. Draw your data (e.g. as a table). 2. Ants will come. 3. Maybe you can read it. 4. Ants will be gone and you can go home.

Community interpretation: Ant Pen makes the impermanence of data immediate, similar to Damien Hurst's decaying shark (Hurst, 1991). In this case, the data points may literally move and depart during authoring: there will be no singular authoritative state of the visualization.

None of the wacky tools reminds the researcher as much of the illusionary stability of data as the Ant Pen. Trying to express a definite visual conclusion about any data will only result in a fleeting impression that evokes the ephemeral, situated, and subjective nature of all data interpretation.

Finally, it also reminds that data and visualization have a cost. The pen requires a disturbing powersource—dead ants—but turns this power into something alive—ants—and possibly meaningful—data. The impermanence of the data—the moving ants—is the very sign that it is alive.

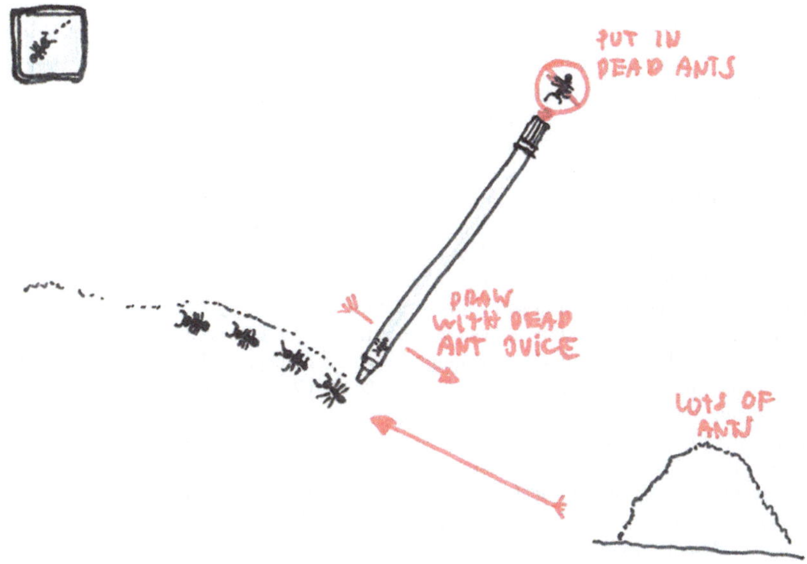

Image 17.11 The Ant Pen

17.4 Summary

The workshop results demonstrated that creative innovation in data visualization can be led by graphic design activity. The vision of tools that had unexpected outcomes suggests many ways of conceptualizing phenomena as they are abstracted into formal procedures. Many of these ideas depended on organic metaphors or processes such as growth or decay and random behaviors such as scattering or dispersal, intermittent or unpredictable motion. Other novelties hint at the ways automated processes could draw on emerging models of machine assisted learning, large language models or image processing. While hand-drawn approaches to data design have enjoyed some popular acclaim (Lupi & Posavec, 2016) as well as having deep historical roots (Friendly, 2006), there have been very few systematic discussions or implementations of these provocative methods. Our contribution has been to demonstrate the rich capacity of imaginative work as a way of broadening the visual tool palette and in the process, expanding the conceptual foundations for data modeling and design.

Acknowledgements This workshop was conceived and performed at the Dagstuhl Seminar on Visualization and the Digital Humanities #23381 2023. https://www.dagstuhl.de/en/seminars/seminar-calendar/seminar-details/23381

References

Apple System 7 Macintosh Basics. www.youtube.com/watch?v=3ScS4OYDfHE

Bertin, J. (1983). *The Semiology of Graphics* (W. Berg, Trans.). University of Wisconsin Press.

Brath, R. (2021). *Visualizing with Text.* CRC Press.

Cameron, W. B. (1963). *Informal Sociology: A Casual Introduction to Sociological Thinking.* Random House.

Cleveland, W. S., & McGill, R. (1985, August). Graphical Perception and Graphical Methods for Analyzing Scientific Data. *Science, 229*(4716), 828–833.

Dealberto, C., Grandin, J., & Schuwey, C. (2022). *L'Atlas Molière.* Les Arènes.

Drucker, J. (2014). *Graphesis.* Harvard University Press.

Friendly, M. 2006. A Brief History of Data Visualization. http://datavis.ca/papers/hbook.pdf

Heer, J., & Bostock, M. (2010). Crowdsourcing Graphical Perception: Using Mechanical Turk to Assess Visualization Design. In *Proceedings of the SIGCHI Conference on Human Factors in Computing Systems.*

Hickman, C. (2019). Kid Pix—The Early Years. www.red-green-blue.com/kid-pix-the-early-years

Hurst, D. (1991). The Physical Impossibility of Death in the Mind of Someone Living. https://en.wikipedia.org/wiki/The_Physical_Impossibility_of_Death_in_the_Mind_of_Someone_Living

Koskinen, I., Zimmerman, J., Binder, T., Redstrom, J., & Wensveen, S. (2011). *Design Research Through Practice.* Elsevier.

Lupi, G., & Posavec, S. (2016). *Dear Data.* Princeton Architectural Press.

Moynihan, B., & Putra, A. (2019). Prototyping the Archival Ephemeral: Experimental Interfaces for the Edwin Morgan Scrapbooks. *Digital Studies/Le champ numérique, 9*(1), 1–36.

Offenhuber, D. (2023). *Autographic Design.* MIT Press.

Ware, C. (2000). *Information Visualization: Perception for Design.* Morgan Kaufmann.

Wilkinson, L. (1999). *The Grammar of Graphics.* Springer.

MIX
Papier aus verantwortungsvollen Quellen
Paper from responsible sources
FSC® C105338

If you have any concerns about our products,
you can contact us on
ProductSafety@springernature.com

In case Publisher is established outside the EU,
the EU authorized representative is:
Springer Nature Customer Service Center GmbH
Europaplatz 3, 69115 Heidelberg, Germany

Printed by Libri Plureos GmbH
in Hamburg, Germany